SpringerBriefs in Electrical and Computer Engineering

T0214349

More information about this series at http://www.springer.com/series/10059

Danda B. Rawat • Min Song • Sachin Shetty

Dynamic Spectrum Access for Wireless Networks

 Springer

Danda B. Rawat
Department of Electrical Engineering
Georgia Southern University
Statesboro, GA, USA

Min Song
Department of Computer Science
Michigan Technological University
Houghton, MI, USA

Sachin Shetty
Department of Electrical
    and Computer Engineering
Tennessee State University
Nashville, TN, USA

ISSN 2191-8112   ISSN 2191-8120   (electronic)
SpringerBriefs in Electrical and Computer Engineering
ISBN 978-3-319-15298-1   ISBN 978-3-319-15299-8   (eBook)
DOI 10.1007/978-3-319-15299-8

Library of Congress Control Number: 2015931510

Springer Cham Heidelberg New York Dordrecht London

Printed on acid-free paper

Springer International Publishing AG Switzerland is part of Springer Science+Business Media
(www.springer.com)

To Our Families

Preface

Wireless communication is the fastest growing sector of the communication industry because of the invention of smartphones and successful deployment of Wi-Fi and cellular networks. Users have anywhere, anytime connectivity to the Internet using their devices. In the Internet of Things (IoT) era, with the increasing number of subscriptions, wireless traffic is increasing exponentially. Spectrum scarcity problem created by static radio frequency (RF) allocation policy will worsen when billions of devices are connected to the Internet. Thus, dynamic spectrum access using software-defined radios (aka cognitive radio networks) could solve the artificial spectrum scarcity problem by allowing unlicensed users to access licensed bands opportunistically. To fully realize dynamic spectrum access in cognitive radio networks, there are policy level as well as technical challenges. Yet, it is very challenging to adaptively allocate resources for dynamic spectrum access in cognitive radio networks for unlicensed secondary users while protecting primary users from any harmful interference.

For dynamic spectrum access, secondary users must identify spectrum opportunities (by sensing channels or searching them in a database) to use them dynamically. Once spectrum opportunities are identified, secondary users switch to transmission mode to communicate with their receivers. This book presents a brief survey of spectrum sensing techniques that are used to find idle RF bands and different approaches for adaptive resource allocation in spectrum underlay and spectrum overlay cognitive radio networks for infrastructure-based and (infrastructure-less) peer-to-peer-based communications. This book is organized as follows:

- Chapter 1 presents an overview of cognitive radio networks, cognitive cycle, and spectrum sensing techniques.
- Chapter 2 presents resource allocation for spectrum underlay cognitive radio networks.
- Chapter 3 presents cloud-integrated resource allocation for spectrum overlay cognitive radio networks.

- Chapter 4 presents distributed computing for cloud-integrated resource allocation for dynamic spectrum access in cognitive radio networks for infrastructure-based and infrastructure-less communications.
- Chapter 5 presents dynamic spectrum access for vehicular network users with cognitive radios.

Statesboro, GA, USA   Danda B. Rawat
Houghton, MI, USA   Min Song
Nashville, TN, USA   Sachin Shetty

Acknowledgments

We would like to express our warm appreciation to Georgia Southern University, Michigan Technological University, and Tennessee State University. We would like to thank Professor Xuemin "Sherman" Shen and the Springer staff who helped us to publish our work. Special thanks go to U.S. National Science Foundation and Department of Homeland Security. Last but certainly not least, we want to thank our families, who never stopped supporting and encouraging us.

Statesboro, GA, USA   Danda B. Rawat
Houghton, MI, USA   Min Song
Nashville, TN, USA   Sachin Shetty

Contents

List of Figures

List of Tables

Chapter 1
An Overview of Cognitive Radio Networks

1.1 Introduction

Wireless communication is the fastest growing segment of the communication industry. With the successful deployment of cellular networks in licensed bands and Wi-Fi networks in unlicensed bands, users have anytime, anywhere connectivity with the networked systems leading to the Internet of Things (IoT). Traditional wireless networks rely on static spectrum assignment where the government regulatory bodies, such as the Federal Communication Commission (FCC) in the United States, assign the Radio Frequency (RF) spectrum to the service providers in an exclusive manner for long term and vast geographic area. Most of the usable RF spectrum are already assigned to certain services leaving no bands for further development of new wireless systems. Furthermore, when everything (such as refrigerator, microwave oven, smart car, etc) is connected to internet, this scarcity would be more severe. However, recent studies show that the static RF spectrum assignment leads to inefficient use of RF spectrum since most of the channels are used only from 15 to 85 % or idle most of the time [2, 14, 22]. Thus the bottleneck created is not because of lack of RF spectrum but because of wasteful static assignments of RF spectrum for long-time and vast geographic area.

Lately, dynamic spectrum access (DSA) in cognitive radio networks (CRNs) using software defined radio (SDR) and machine learning has been proposed to facilitate the RF spectrum sharing so that the larger density of wireless devices can be accommodated without needing new RF spectrum bands. For DSA in CRNs, licensed incumbent users are regarded as primary users (PUs) and unlicensed users are secondary users (SUs) who do not have their own licensed spectrum but have capability of accessing others' RF spectrum in an opportunistic manner. In CRNs, SUs are allowed to use the spectrum opportunities dynamically without creating harmful interference to PUs. Thus, in order not to interfere PUs, SUs must be able to detect PU signals quickly and efficiently.

© The Author(s) 2015
D.B. Rawat et al., *Dynamic Spectrum Access for Wireless Networks*, SpringerBriefs
in Electrical and Computer Engineering, DOI 10.1007/978-3-319-15299-8_1

Typically, SUs are invisible to PUs in CRN, therefore possibly there are no changes needed for PU devices. SUs are either allowed to transmit with low transmit power such as in ultra-wide band (UWB) technology or allowed to access idle spectrum bands dynamically without creating any harmful interference to PUs. The former approach is known as spectrum underlay as shown in as Fig. 1.2b where SUs and PUs coexist while accessing bands simultaneously. The latter one is known as spectrum overlay as shown in as Fig. 1.2a where SUs identify idle bands and use them opportunistically so as not to interfere PUs [14, 19, 33].

1.2   Cognitive Radio Networks

In CRN, SUs access spectrum bands which is not licensed to them opportunistically without interfering licensed PUs. Thus, to find RF spectrum opportunities, individual SUs undergo through a cognitive cycle [21] as shown in Fig. 1.1. For DSA in CRN [2, 13–15], there are three basic phases that SUs go through:

- Sensing phase (aka observe phase): During this phase, SUs sense channels for spectrum occupancy information using different algorithms to identify spectrum opportunities by using analysis (aka reasoning);
- Adaptation phase (aka switching phase): During this phase SUs switch from sensing phase (receiving mode) to transmit mode for actual communications;
- Act phase (aka communication phase): During this phase, SU communicates to its receiver based on newly adapted transmit parameters.

Typical spectrum sensing techniques are discussed in the following sections.

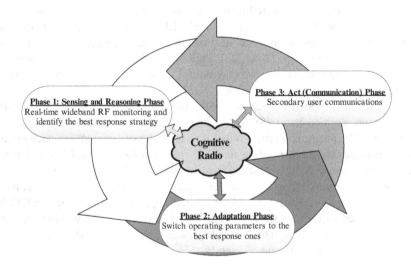

Fig. 1.1 Cognitive cycle for cognitive radio network

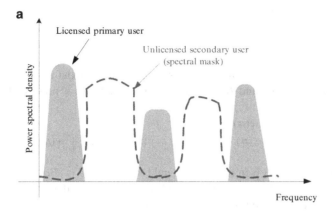

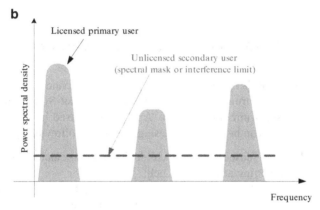

Fig. 1.2 Spectrum overlay and spectrum underlay approaches for dynamic spectrum access in CRN. (**a**) Spectrum overlay. (**b**) Spectrum underlay

1.3   System Model for Cognitive Radio Networks

As shown in Fig. 1.1, each SU goes through three different stages. During sensing phase, the received signal $y(t)$ at SU in continuous time can be expressed as

$$y(t) = gs(t) + w(t) \tag{1.1}$$

where h is channel gain between PU transmitter and SU receiver, $s(t)$ is the PU's signal (to be detected), and $w(t)$ is the additive Gaussian white noise (AWGN).

We consider that the signal has central frequency f_c and bandwidth W, and has sampled the received signal at a sampling rate f_s with $f_s > W$, and the sampling period $T_s = 1/f_s$. Thus, we can write the sampled received signal for (1.1) as

$$y(n) = gs(n) + w(n) \tag{1.2}$$

For an ideal channel i.e., $g = 1$, the received signal is

$$y(n) = s(n) + w(n) \qquad (1.3)$$

Based on the received signal, two possible hypotheses exist as: $\mathcal{H}_0$ to represent the signal $s(n)$ is absent, and $\mathcal{H}_1$ to represent the signal $s(n)$ is present. That is,

$$\begin{aligned}
\mathcal{H}_0 &: \quad y(n) = w(n) \\
\mathcal{H}_1 &: \quad y(n) = s(n) + w(n) \quad \text{or} \quad y(n) = gs(n) + w(n)
\end{aligned} \qquad (1.4)$$

When $s(n) = 0$, the given frequency band is idle and SU is allowed to access the given band. When $s(n) \neq 0$, the given band is not idle (if the detection is error free) and no SUs are allowed to access the given band.

1.4 Spectrum Sensing in Cognitive Radio Networks

Spectrum sensing in CRNs is fundamental step to find idle bands. Once SUs sense the spectrum bands, spectrum analysis is carried out so that SUs make informed decision about any spectrum opportunities. Analysis of sensed information to find spectrum opportunities can be carried out using one of the following approaches.

- SUs can analyze the sensed data to make decision about channel occupancy on their own without getting any external help.
- SUs can collaboratively communicate with their peers to exchange their sensed data and analyze the received data in a distributed manner to find idle channels.
- SUs, who do not have sensing capabilities, can receive information about idle channel from external agent that has global channel occupancy information.
- SUs sense and report their sensed information to the fusion center. The fusion center analyzes the data for SUs to identify idle bands and notifies to the SUs.

There are several approaches for spectrum sensing and analysis in CRNs which are discussed briefly in the following section [14, 19, 20, 31].

1.4.1 Primary Transmitter Detection

In a primary transmitter detection approach, SU who wants to use spectrum opportunities overhears the PU signal and uses one of the following approaches to detect the signal locally for given time and location.

- Matched Filtering Based Signal Detection: Matched Filtering (MF) based spectrum sensing is an optimal method for detection of PUs [17] as it maximizes the signal-to-noise ratio (SNR) corresponding to the Eq. (1.3)

$$\gamma = \frac{|s(n)|^2}{E[w^2(n)]} \tag{1.5}$$

However, MF is not a good choice for CRNs since it requires perfect knowledge of the PU signal (such as modulation, frequency, etc.) and it is complex to implement for wide band regime [6]. Furthermore, huge amount of power is consumed to scan wide range of frequency bands.

- Covariance Based Signal Detection: The covariance based signal detection method [32] exploits the covariances of signal and noise since the statistical covariances of signal and noise are usually different. This covariance property helps to differentiate signal from noise where the sample covariance matrix of the received signal is computed at the receiver [32]. The received signal (1.2) can be represented in a vector form. Then the covariance matrix of received signal is computed. When the given channel is idle, the covariance matrix of the received signal is equal to 0. When there is a signal in a given channel, signal samples are correlated and covariance matrix is not equal to 0. Thus, by using covariance matrix of the received signal, SUs can detect the presence of PU signal in a given channel.

- Waveform-Based Detection: Synchronous communication system uses patterns corresponding to a transmitted signal such as preambles, midambles, regularly transmitted pilot patterns, spreading sequences, etc. This information could be used to detect the presence of the signal. The received signal could be compared with a known copy of itself to detect the presence of PU signal and the process is known as waveform-based detection [23]. For a received signal (1.3), the detection metric for waveform-based detection can be computed as

$$D = Re \left[\sum_{n=1}^{N} y(n)s^*(n) \right] \tag{1.6}$$

where N is length of known pattern. When the PU signal is absent (present), the D will have only noise times copy of the signal (noise plus signal times copy of the signal). Then, D can be compared with a given threshold λ_T to detect whether there is any signal or not. We can compute the probability of true detection P_T (i.e. successful detection when signal is present in the channel) and the probability of false alarm P_F (i.e. the detection indicated that the channel is occupied however actually it is not) as

$$P_T = Pr(D > \lambda_T | \mathcal{H}_1) \quad , \quad P_F = Pr(D > \lambda_T | \mathcal{H}_0) \tag{1.7}$$

The main goal of any signal detection techniques is to reduce P_F and increase P_T. The performance of the waveform based detection depends on chosen threshold λ_T and it can be predicted based on noise variance. The waveform-based detection requires short detection time but is susceptible to synchronization errors [7].

- Energy Based Detection: Energy based signal detection is regarded as efficient spectrum sensing approach as it has low computational and implementation complexities [6]. In energy based detection, the SU receiver doesn't need any knowledge of PU signal. This approach compares the energy of the received signal with a given threshold [26]. Note that the fixed threshold could lead to inaccurate signal detection thus the threshold value should be adaptive to the noise floor. For a system model given in (1.3), the energy of the received signal can be computed as

$$D = \sum_{n=0}^{N} |y(n)|^2 \tag{1.8}$$

Then, decision can be made using two hypotheses. The false alarm and true detection probabilities are calculated by comparing computing energy with the threshold as in (1.7). In energy based detection, there are some disadvantages such as the choice of threshold, poor performance under low SNR and inability to differentiate interference (e.g., in CDMA) and noise.

- Cyclostationary Based Detection: This approach takes advantage of cyclo-stationary properties of the received signals at SUs to detect PU transmissions [10, 16]. Typical transmitted PU signals are stationary random process however due to modulation of signals with sinusoid carriers, cyclic prefix in OFDM, etc., cyclostationarity features (such as periodicity in mean and autocorrelation) are induced. Furthermore, the noise is Wide-Sense Stationary (WSS) with no correlation. Therefore, cyclostationary based approach can differentiate PU signals from noise [16]. The cyclic spectral correlation function (SCF) of the received signal is used to detect whether the given channel is idle or nor. The cyclic SCF of (1.3) is expressed as

$$S_{yy}^{\alpha}(f) = \sum_{\tau=-\infty}^{\infty} R_{yy}^{\alpha}(\tau)e^{-j2\pi f} \tag{1.9}$$

where $R_{yy}^{\alpha}(\tau)$ is the cyclic autocorrelation function and is periodic in n, and α is the cyclic frequency. Note that when $\alpha = 0$, the SCF becomes power spectral density. Peak is present in cyclic SCF implies that the PU signal is present. When there is no such peak, it indicates that the given channel is idle at given time and location.

- Random Hough Transform Based Detection: Random Hough transform, a method used in image pattern recognition, has been proposed for PU signal detection [8]. In this method, Random Hough transform of received signal $y(n)$ is used to detect the presence of radar pulses using known patterns in IEEE 802.11 wireless channels.

- Radio Identification Based Detection: Wireless features such as modulation type, channel, etc. can be extracted from the received signal $y(n)$ in (1.3) at SU. These features are used by SUs to select the best parameters for them to transmit their information while protecting PUs [9].
- Wavelet Based Detection: Wavelet based signal detection, a popular method in image processing for edge detection, has been proposed for signal detection in CRN [25]. This approach is used to detect edges in the power spectral density (PSD) of different channels. This approach detects the edges in PSD of non-overlapping sub-bands to find idle spectrum opportunities in given location and time.
- Multi-Taper Spectrum Sensing: In Multi Taper spectrum estimation (MTSE), SU receiver collects last N samples in a vector form known as a set of Slepian base vectors [13, 24]. The Fourier transform of Slepian vectors gives the maximal energy concentration in the bandwidth $f_c - W$ to $f_c + W$ when there is PU signal. Based on this information, SUs can identify the idle spectrum bands and utilize them dynamically.

1.4.2   Primary Receiver Detection

Alternatively, SUs could use PU receiver detection within their communication range to identify whether the given channel is idle or not. Typical PU receiver emits the local oscillator (LO) leakage power from its RF front-end while receiving the data from its PU transmitter. SU uses that leakage power to detect the PU signal [28]. SUs need some sensors to detect the LO leakage and analyze the sensed data to identify whether the given channel is busy for given time and location. In this approach, SUs need extra sensors to detect local oscillator leakage power but PUs do not need any modification.

1.4.3   Cooperative Detection

Individual SUs can find the spectrum opportunities locally based on their own sensed information. However, decision made by a single wireless device could mislead the detection process. Thus, signal detection by collaborating among different SUs could improve the overall performance and accurate information regarding spectrum opportunities [7]. Collaborative method helps to minimize probability of miss-detection and false alarms, and to solve the hidden PU problem [7]. Different approaches for spectrum sensing using collaborative detection is presented below.

- Centralized Server Based Cooperative Detection: In this approach, a central unit (aka a fusion center or server) collects sensed information from SUs and aggregates the information to find spectrum opportunities [30]. Then, the server broadcasts the information of spectrum status to all SUs. Based on the received information, SUs adapt their transmit parameters to communicate to their receivers. In centralized server based detection, the server does not have the capability of sensing the spectrum. Alternatively, external detection technique has been proposed where SUs do not have sensing capabilities but receive spectrum occupancy information from a server which has sensing capability using external detection agent [12]. There are few advantages of using this approach as it helps to overcome hidden PU terminal problem, helps to reduce uncertainty due to shadowing and fading [12], and helps to minimize the power consumptions for sensing and signal detection [12].
- Distributed Cooperative Approach: Unlike centralized and external detection approaches, SUs interact with other peers to collaboratively share their sensed information in distributed cooperative approach. Then, SUs use that information to identify idle spectrum using algorithms such as spectrum load smoothing algorithms [5].

1.4.4 Interference Temperature Management

In spectrum underlay approach, SUs are allowed to coexist and share the spectrum with PUs as long as they do not exceed the upper interference limit for given frequency band and geographic location [4, 29] as shown in Fig. 1.2b. In this case, SUs must be able to estimate about how much interference they are creating to PUs in a given channel so as not the exceed the specified threshold at PUs. As noted, in spectrum underlay approach, SUs do not have to sense to identify the idle spectrum bands, however, the SUs can not use high transmit power even though the licensed system is completely idle because of imposed low transmit power and interference temperature limit at PUs.

1.5 Adaptation and Act/Communication Phases

In case of a SU with one radio for sensing and communication, the SU switches the mode from sensing to transmit mode to communicate with its receiver using suitable transmit parameters. In the case of a SU with two radios (one for sensing and another for actual communications), the SU adapts its transmit parameters such as power, rate, modulations techniques, etc., that are suitable to a given channel for communications. It is important to choose suitable parameters for DSA in CRNs for reliable SU communications.

1.6 Challenges and Motivations

With the successful deployment of cellular Wireless systems and Wi-Fi network, wireless users and data traffic is growing exponentially [1]. Operators of licensed bands do not allow SUs to access their bands opportunistically. As a consequence, severe spectrum shortage will be faced in IoT era [2, 3, 11, 14, 27] where almost all objects will be connected to the Internet and controlled by the networked systems [3, 18].

Dynamic spectrum access in CRN has been proposed to alleviate the spectrum scarcity problem created by static spectrum allocation [2, 13, 14, 22]. However, there are more challenges to be addressed to fully realize CRNs. Some of them are listed below

- How quick and reliable is the spectrum sensing and analysis process to identify idle bands in wide band regime? Is spectrum sensing and detection process is quick enough not to interfere any PU communications?
- How much interference created by SUs is tolerable by PUs? How long PUs can tolerate certain level of interference once SUs detect PUs in their licensed bands?
- Is DSA in CRN secure enough to convince primary service providers or PUs to leverage their licensed band opportunistically?
- Are licensed service providers willing to give their licensed bands to SUs for free?
- How fast SUs are leaving the licensed bands once PUs are back to use their licensed bands?
- What benefit can be received by PUs or service providers?
- Can wireless devices used by SUs sense and process data for wide band regime? How fast they can process that data?
- Can SUs leverage cloud computing platform for processing big data?

We have presented the materials to address these questions to some extent in the remaining chapters.

1.7 Organizations and Summary

Dynamic spectrum access in cognitive radio networks is regarded as a solution to spectrum scarcity problem to handle billions of wireless devices in the IoT era. This book presents dynamic spectrum access in both spectrum underlay and spectrum overlay cognitive radio networks where licensed SUs access spectrum bands that are licensed to PUs in an opportunistic manner without creating any harmful interference to PUs while meeting their own QoS requirements.

Chapter 2 presents resource allocation for spectrum underlay in CRN where SUs coexist and share spectrum with PUs. SUs make sure that they do not create any harmful interference through admissibility check, power constraint and interference constraint to protect PUs. Chapter 3 presents resource allocation in spectrum overlay in CRN where SUs avoid channels used by PUs. A two stage based Stackelberg game is proposed for spectrum servers and SUs where SUs maximize their rates and spectrum servers maximize their payoff subject to their respective constraints. Chapter 4 presents geolocation-aware resource allocation for SUs in CRNs where most of the computing, storing and searching process is handled by the cloud computing platform using Storm model and Cassandra database. Once SUs find spectrum opportunity for given location and time, they can use those opportunities to communicate with their receivers using either infrastructure based communications or peer-to-peer based communications. Chapter 5 presents resource allocation for SUs in cognitive radio enabled vehicular networks where SUs search idle spectrum in a spectrum database. Analysis for network connectivity for vehicles is presented which depends on vehicle density, speed of the vehicles, and transmission range. The performance of algorithms is evaluated using numerical results obtained from simulations.

References

1. Cisco, 2013: Cisco Visual Networking Index: Global Mobile Data Traffic Forecast Update, 2012–2017.   http://www.cisco.com/en/US/solutions/collateral/ns341/ns525/ns537/ns705/ns827/white_paper_c11-520862.pdf.
2. I. F. Akyildiz, W.-Y. Lee, M. C. Vuran, and S. Mohanty. NeXt Generation/Dynamic Spectrum Access/Cognitive Radio Wireless Networks: A Survey. *Computer Networks*, 50(13):13–18, September 2006.
3. Luigi Atzori, Antonio Iera, and Giacomo Morabito. The internet of things: A survey. *Computer networks*, 54(15):2787–2805, 2010.
4. J. Bater, Hwee-Pink Tan, K.N. Brown, and L. Doyle. Modelling Interference Temperature Constraints for Spectrum Access in Cognitive Radio Networksr. In *Proceeding of IEEE International Conference on Communications, 2007, ICC'07*, pages 6493–6498, June 2007.
5. Lars Berlemann, Stefan Mangold, Guido R. Hiertz, and Bernhard H. Walke. Spectrum Load Smoothing: Distributed Qquality-of-Service Support for Cognitive Radios in Open Spectrum. *European Transactions on Telecommunications*, 17:395–406, 2006.
6. D. Cabric, S. Mishra, and R. Brodersen. Implementation Issues in Spectrum Sensing for Cognitive Radios. In *Asilomar Conf. on Signals, Systems and Computers*, pages 772–776, Pacific Grove, California, June 2004.
7. D. Cabric, A. Tkachenko, and R. Brodersen. Spectrum Sensing Measurements of Pilot, Energy, and Collaborative Detection. In *Proceedings IEEE Military Commun. Conf.*, pages 1–7, October 2006.
8. K. Challapali, S. Mangold, and Z. Zhong. Spectrum Agile Radio: Detecting Spectrum Opportunities. In *Proc. Int. Symposium on Advanced Radio Technologie*, Boulder, CO, Mar. 2004.
9. T. Farnham, G. Clemo, R. Haines, E. Seidel, A. Benamar, S. Billington, N. Greco, N. Drew, B. Arram T. Le, and P. Mangold. IST-TRUST : A Perspective on the Reconfiguration of Future Mobile Terminals using Software Download. In *Proc. IEEE Int. Symposium on Personal, Indoor and Mobile Radio Commun.*, pages 1054–1059, London, UK, September 2000.
10. W. Gardner. Exploitation of Spectral Rredundancy in Cyclostationary Signals. *IEEE Signal Processing Mag.*, 8(2):14–36, 1991.

11. David Goldman. Sorry, America: Your Wireless Airwaves are Full. February 21, 2012. http:// money.cnn.com/2012/02/21/technology/spectrum_crunch/index.htm.
12. Z. Han, R. Fan, and H. Jiang. Replacement of Spectrum Sensing in Cognitive Radio. *IEEE Transactions on Wireless Communications*, 8(6):2819–2826, June 2009.
13. S. Haykin. Cognitive Radio: Brain-Empowered Wireless Communications. *IEEE J. Select. Areas Commun.*, 3(2):201–220, Feb 2005.
14. Jaime Lloret Mauri, Kayhan Zrar Ghafoor, Danda B. Rawat, and Javier Manuel Aguiar Perez. *Cognitive Networks: Applications and Deployments*. CRC Press, 2014.
15. J. Mitola and G. Q. Maguire. Cognitive Radio: Making Software Radios More Personal. *IEEE Personal Communications Magazine*, 6(6):13–18, August 1999.
16. M. Öner and F. Jondral. Air Interface Identification for Software Radio Systems. *AEÜ International Journal of Electronics and Communications*, 61(2):104–117, February 2008.
17. J. G. Proakis. *Digital Communications*. McGraw Hill, Boston, MA, fourth edition, 2000.
18. Danda B Rawat, Joel Rodrigues, and Ivan Stojmenovic. *Cyber Physical Systems: From Theory to Practice*. CRC Press, USA, 2015.
19. Danda B Rawat and Gongjun Yan. Signal processing techniques for spectrum sensing in cognitive radio systems: Challenges and perspectives. In *IEEE AH-ICI 2009 Conference*, pages 1–5, 2009.
20. Danda B Rawat and Gongjun Yan. Spectrum sensing methods and dynamic spectrum sharing in cognitive radio networks: A survey. *International Journal of Research and Reviews in Wireless Sensor Networks*, 1(1):1–13, 2011.
21. Rajesh Sharma and Danda B. Rawat. Advances on Security Threats and Countermeasures, for Cognitive Radio Networks: A Survey. *IEEE Communications Surveys and Tutorials*, 2015. DOI: 10.1109/COMST.2014.2380998.
22. Min Song, Chunsheng Xin, Yanxiao Zhao, and Xiuzhen Cheng. Dynamic spectrum access: from cognitive radio to network radio. *IEEE Wireless Communications*, 19(1):23–29, 2012.
23. H. Tang. Some Physical Layer Issues of Wide-band Cognitive Radio Systems. In *IEEE Int. Symposium on New Frontiers in Dynamic Spectrum Access Networks*, pages 151–159, Baltimore, MD, June 2005.
24. D. J. Thomson. Spectrum Estimation and Harmonic Analysis. *Proc. IEEE*, 20:1055–1096, Sep. 1982.
25. Z. Tian and G. B. Giannakis. A Wavelet Approach to Wideband Spectrum Sensing for Cognitive Radios. In *Proc. IEEE Int. Conf. Cognitive Radio Oriented Wireless Networks and Commun. (Crowncom)*, pages 1054–1059, Mykonos, Greece, June 2006.
26. H. Urkowitz. Energy Detection of Unknown Deterministic Signals. In *Proceedings of the IEEE*, volume 55, pages 523–531, April 1967.
27. Rolf H Weber and Romana Weber. *Internet of Things*. Springer, 2010.
28. B. Wild and K. Ramchandran. Detecting Primary Receivers for Cognitive Radio Applications. In *proceeding of IEEE Dynamic Spectrum Access Networks, DySPAN 2005*, pages 124–130, November 2005.
29. Yiping Xing, Chetan N. Mathur, M.A. Haleem, R. Chandramouli, and K.P. Subbalakshmi. Dynamic spectrum access with qos and interference temperature constraints. *IEEE Transactions on Mobile Computing*, 6(4):423–433, 2007.
30. R.D. Yates, C. Raman, and N.B. Mandayam. Fair and Efficient Scheduling of Variable Rate Links via a Spectrum Server. In *Proceeding of IEEE International Conference on Communications, 2006, ICC'06*, pages 5246–5251, June 2006.
31. Tevfik Yucek and Hüseyin Arslan. A survey of spectrum sensing algorithms for cognitive radio applications. *Communications Surveys & Tutorials, IEEE*, 11(1):116–130, 2009.
32. Yonghong Zeng and Ying-Chang Liang. Spectrum-Sensing Algorithms for Cognitive Radio Based on Statistical Covariances. *IEEE Transactions on Vehicular Technology*, 58(4): 1804–1815, May 2009.
33. Qing Zhao and B. M. Sadler. A Survey of Dynamic Spectrum Access. *IEEE Signal Processing Magazine*, 24(3):79–89, May 2007.

Chapter 2
Resource Allocation in Spectrum Underlay Cognitive Radio Networks

2.1 Overview

In spectrum overlay approach, SUs coexists with PUs and share their spectrum as shown in Fig. 1.2b. However, each SU has imposed transmit power constraint so as not to create any harmful interference to active PUs. The main goal of each SU is to maintain lower interference level than the specified tolerable level at PUs while sharing PUs' licensed bands dynamically. To maintain low interference level, SUs must transmit with lower transmit power. Thus power control is essential for each SUs. In wireless communications, power control is performed to satisfy the specified minimum signal-to-interference-plus-noise (SINR) to get desired data rate [2, 7]. For voice communications, once target SINR level is met, there will be no improvement in voice quality by increasing power or SINR. However, for data communications, increase in SINR results in increase in data rate. Thus, SUs try to increases their SINR values by increasing their transmission powers while satisfying their imposed power constraints in data communications.

In future wireless systems, wireless devices will be used for both data and voice communications. Thus, the approach that provides optimal values of SINR and power is more suitable for future applications. Note that when we deal with the SINR greater than or equal to target SINR, we need to optimize over entire feasible SINR region which leads the optimization problem to non-convex. One way to solve this type of problem is to consider the throughput $\sum_i w_i \log_2(1 + \text{SINR}_i)$ maximization for SUs for given bandwidth w_i and $SINR_i$.

Since SUs non-cooperatively maximize their rates or throughput using their optimal powers, the rate and power adaptations of SUs can be formulated as a non-cooperative game. In this game, payoff or surplus of the SUs can be maximized using non-cooperative strategies.

In spectrum underlay approach, we apply non-cooperative game for power adaptation for active SUs to satisfy imposed quality-of-service (QoS) constraints on them and interference power constraints at PUs to maximize overall payoff for SUs

© The Author(s) 2015   13
D.B. Rawat et al., *Dynamic Spectrum Access for Wireless Networks*, SpringerBriefs
in Electrical and Computer Engineering, DOI 10.1007/978-3-319-15299-8_2

using distributed interference compensation. Spectrum sharing in similar scenario has been performed in [4, 5] where PUs and SUs are treated equally in [4], and readmission control and distributed interference compensation is not considered in [5].

2.2   Network Model and Problem Formulation

A communication network model for spectrum underlay spectrum sharing is shown in Fig. 2.1 where a primary base station communicates with M PUs using licensed bandwidth W. There are N SU pairs who share the same licensed bandwidth W and form N links for SU communication using code division multiple access technology.

The SINR corresponding to a given link i can be written as

$$\gamma_i = \frac{W}{R_i} \frac{g_{i,i}^{(s)} p_i}{\sum\limits_{j=1, j \neq i}^{N} g_{i,j}^{(s)} p_j + N_i}, \qquad \forall i = 1, \ldots, N \qquad (2.1)$$

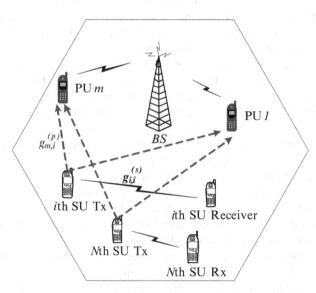

Fig. 2.1 Network model with two PUs communicating with their base station and $N = 2$ SU links

Where

- a link gain between SU transmitter j and SU-receiver i is $g_{i,j}^{(s)}$, and a link gain between SU transmitter j and PU receiver i pair is $g_{i,j}^{(p)}$. Note that channel gains are inversely proportional to $d_{i,j}^4$ for $d_{i,j}$ distance between transmitter j and receiver i.
- $K_i = W/R_i$ is processing gain, and
- N_i is the interference from active PUs, i.e.,

$$N_i = \sum_{m=1}^{M} \tilde{\alpha}_m g_{i,m}^{(p)} p_m + n_i, \qquad \forall i = 1, \ldots, N$$

where n_i is additive white Gaussian noise, and $\tilde{\alpha}_m \in \{0, 1\}$ denotes $\{$OFF, ON$\}$ status of PU m.

From PU's perspective, SUs are invisible and active PUs are assumed to be capable of tolerating a maximum interference power p_m^{th}. This tolerable power level should be supplied by the government regulatory bodies which is discussed in Chap. 1 (Sect. 1.4.4). In order to protect PU communications, the interference created by SUs to all PUs $m = 1, \ldots, M$ should satisfy the following constraint

$$\sum_{i=1}^{N} g_{m,i}^{(p)} p_i \leq p_m^{th}, \qquad \forall m = 1, \ldots, M \tag{2.2}$$

All SUs are assumed to be aware of the total interference caused to PUs and the PUs' interference threshold p_m^{th}, $\forall m$ where PUs broadcasts the related information or SUs use extra sensors to estimate the interference.

Furthermore, each SU needs to satisfy its minimum required SINR $\overline{\gamma}_i$ to meet its rate or QoS requirement. That is, each SU should satisfy the following constrain for rate or QoS requirements.

$$\gamma_i \geq \overline{\gamma}_i, \qquad \forall i = 1, \ldots, N \tag{2.3}$$

Note that SUs who cannot satisfy QoS constraint in (2.3) are subject to drop from the system or to transmit with different rates R_i using different processing gain K_i or different modulation techniques. When, for a given SU, there are many rates that satisfy the QoS/rate requirements, SU will pick up the highest rate among feasible rates that satisfies (2.3).

2.2.1 Distributed Admission Control for SUs

Admission control helps to limit the number of SUs for dynamic spectrum access which helps active SUs to have reliable communications. Admission control also boosts the overall network performance by dropping users who create interference and do not meet their own QoS/rate requirements. Furthermore, admission control blocks the new SU requests who do not meet the admission criteria based on their locations. The former one is known as post admission control and latter one is known as pre-admission control.

For *pre-admission control*, SUs check their admissibility condition before they start their actual transmission. The pre-admission control criteria [3, 9] for distributed network model can be defined as channel gain ratio as

$$\lambda_i = \frac{g_{i,i}^{(s)}}{\max_{k=1...,M} g_{i,k}^{(p)}} \tag{2.4}$$

Before actual data transmission, SU checks the channel gain ratio λ_i in (2.4) against a given threshold $\lambda_T \geq 0$. If any PU is closer than the SU's receiver, the SU will not be allowed to access the channel as it create the high interference to the PUs. Note that the channel gains between wireless transmitter and receiver can be estimated with the help of some ranging capability.

For *post-admission control*, active SUs check their QoS constraint in (2.3) and interference constraint in (2.2). When PUs resume their communications and start using their licensed channels actively, there will be less spectrum available for SUs. Thus, the SUs who are active in the system but do not satisfy the constraints (2.3) and (2.2) are dropped from the system.

2.2.2 Power Control

Power control in spectrum underlay minimizes interference to other active users and increases the battery life of the devices. Optimal power for a given SU satisfies both interference constraint in (2.2) and SINR requirement in (2.3). To find the optimal power, we substitute γ_i from (2.1) into (2.3) as

$$K_i \frac{g_{i,i}^{(s)} p_i}{\sum_{j=1,j \neq i}^{N} g_{i,j}^{(s)} p_j + N_i} \geq \overline{\gamma}_i \tag{2.5}$$

We can rearrange (2.5) as

$$p_i - \frac{\overline{\gamma}_i \sum\limits_{j=1, j \neq i}^{N} g_{i,j}^{(s)} p_j}{K_i g_{i,i}^{(s)}} \geq \frac{\overline{\gamma}_i N_i}{K_i g_{i,i}^{(s)}}. \tag{2.6}$$

Then, considering equality sign, we can compute power of ith SU link for the next time instant $t + 1$ as

$$p_i(t + 1) = [\mathbf{F}]_{i,j} \, p_i(t) + u_i, \tag{2.7}$$

where $[\mathbf{F}]_{i,j} = \frac{\overline{\gamma}_i g_{i,j}^{(s)}}{K_i g_{i,i}^{(s)}}$, for $i \neq j$ and $[\mathbf{F}]_{i,j} = 0$, for $i = j$. and $u_i = \frac{\overline{\gamma}_i N_i}{K_i g_{i,i}^{(s)}}$. For ith SU link, the optimal transmit power that does not exceed the maximum allowed power limit p_i^{sup} and satisfy the QoS requirements in terms of minimum SINR can be written as

$$p_i(t + 1) = \min \left\{ p_i(t) \frac{\overline{\gamma}_i}{\gamma_i(t)}, p_i^{sup} \right\}. \tag{2.8}$$

2.2.3 Problem Formulation

In spectrum underlay spectrum sharing, constraint optimization problem for joint pre-admission control, power and rate adaptation, and QoS requirements is formulated as

$$\max_{\gamma_i} \sum_{i=1}^{N} \alpha_i \log_2(1 + \gamma_i) \text{ subject to } \begin{cases} p_i^{min} \leq p_i \leq p_i^{sup}, & \forall i = 1, .., N \\ \sum\limits_{i=1}^{N} \alpha_i g_{m,i}^{(p)} p_i \leq p_m^{th}, & \forall m = 1, .., M \\ \gamma_i \geq \overline{\gamma}_i & \forall i = 1, .., N \\ \alpha_i \in \{0, 1\} \end{cases} \tag{2.9}$$

where $p_i^{min} \geq 0$ is the minimum power level, and $\alpha_i = 0$ and $\alpha_i = 1$, respectively, imply that the ith SU link is OFF and ON. The constraint optimization problem (2.9) maximizes the achievable data rates of active SUs by using optimal transmit power and satisfying QoS and interference constraints. Note that when the maximum eigenvalue of the matrix $\mathbf{F}$ is larger than 1, the SUs with chosen minimum SINRs $\overline{\gamma}_i$

are not admissible to the systems to find a solution of the optimization problem (2.9) [9]. However, SUs can readjust their SINRs for their QoS requirements to access the spectrum using optimal values obtained by solving (2.9).

2.3 Game Formulation

For distributed spectrum underlay scenario, individual SUs choose their transmit powers to maximize their payoff/surplus without cooperating with other SUs. For this type of strategies, non-cooperative game is implemented for SUs and the game can be defined as

$$G = \{\mathcal{N}, \mathcal{P}_i, \{u_i(.)\}_{i \in \mathcal{N}}\} \tag{2.10}$$

where

- $\mathcal{N} = \{1, \ldots, N\}$ represents players or active SU links,
- $\mathcal{P}_i$ represents power adaptation strategies of players and player chooses its power from its strategy set $\mathcal{P}_i = \{p_i | p_i \in [p_i^{min}, p_i^{sup}]\}$, $\forall i$, and
- $u_i(.)$ is the utility/payoff received based on the adapted strategy for the SU.

In a non-cooperative game, power profile of opponents of the ith player is $\mathbf{p}_{-i} = (p_1, \ldots, p_{i-1}, p_{i+1}, \ldots, p_N)$ and the complete power profile of all SU is $\mathbf{p} = \{p_i, \mathbf{p}_{-i}\}$. When SU selects its power strategy and it does not satisfy the QoS requirement, it has to reduce its minimum required target SINR for its QoS requirement.

The utility of a player (SU) is a logarithmic function as

$$u_i(\gamma_i) = u_i(\gamma_i(p_i, \mathbf{p}_{-i})) = \theta_i \log_2(1 + \gamma_i), \quad [\text{bits/transmission}] \tag{2.11}$$

where θ_i is the weight of the ith SU link with $\sum_i^N \theta_i = 1$ to maintain fairness among SUs and $\log_2(1 + \gamma_i)$ is the achievable rate of ith link. Thus, maximization of (2.11) for SUs results in maximization of their data rates.

SUs should not violate the interference limit (2.2) in spectrum underlay. However, if they exceed the limit, the pricing function is used to penalize the payoff of the SU. Pricing function is defined as

$$c_i(p_i) = \max\left\{0, \frac{\beta}{M}\left(\sum_{m=1}^{M} g_{m,i}^{(p)} p_i - \sum_{m=1}^{M} \omega_i^m p_m^{th}\right)\right\} \tag{2.12}$$

where $\omega_i^m = \frac{g_{m,i}^{(p)}}{\sum_{i=1}^{N} g_{m,i}^{(p)}}$ is the weighted coefficient, and β is constant factor. From (2.12), we can see that if interference constraint (2.2) is not violated, SU will have zero cost.

Then, surplus (aka net benefit/payoff) which is the difference between utility in (2.11) and cost in (2.12). That is,

$$s_i(p_i) = u_i(\gamma_i) - c_i(p_i) \tag{2.13}$$

When the instantaneous SINR is greater than or equal to the minimum required SINR, the optimal power allocation is given by (2.8) and SUs will have reliable link. Note that the SU's utility function is strictly increasing with $p_i^{min} \leq p_i \leq p_i^{sup}$ for fixed $\mathbf{p}_{-i}$. Moreover, price function incorporates the compensation for violating interference constraint (2.2) and we can get rid of constraint (2.2) in (2.9). Thus, the optimization problem (2.9) can be expressed as:

$$\max \sum_{i=1}^{N} \alpha_i s_i(p_i) \text{ subject to } \begin{cases} p_i^{min} \leq p_i \leq p_i^{sup} \\ \gamma_i \geq \bar{\gamma}_i \end{cases} \quad \forall i = 1, .., N \tag{2.14}$$

At equilibrium point, no players would benefit by deviating their strategies unilaterally. This optimal point of the game is known as Nash Equilibrium (NE) of the game [10]. At NE, the game satisfies

$$u_i(p_i', \mathbf{p}_{-i}') \geq u_i(p_i, \mathbf{p}_{-i}'), \quad p_i' = p_i(t+1), \quad p_i' \in \mathcal{P}_i \tag{2.15}$$

To check whether the solution is optimal at NE or not, we compute the first order derivative of $s_i(p_i)$ with respect to p_i as

$$\frac{\partial s_i(p_i)}{\partial p_i} = \frac{\partial u_i(\gamma_i)}{\partial p_i} - \frac{\beta}{M} g_{m,i}^{(p)} \tag{2.16}$$

and the second order derivative of $s_i(p_i)$ with respect to p_j as

$$\frac{\partial^2 s_i(p_i)}{\partial p_i \partial p_j} \leq 0, \quad \forall i \neq j \tag{2.17}$$

This result indicates that there exist unique optimal solution at NE [10] where power is allocated according to (2.8).

2.4 The Algorithm

First, if a SU wants to access the licensed spectrum, it checks its admissibility criteria using (2.4). If it satisfies the criteria for pre-admission control ad there is idle band, it starts its communication with its receiver. Otherwise it waits until it becomes admissible. Active SUs who do not satisfy constraint (2.3) are dropped form the systems [1]. Otherwise they create unnecessary interference to other active users. Based on above analysis, the algorithm is presented below:

Algorithm 2.1 Power, rate and admission control algorithm

---

1. *Initial input*: Randomly placed M PUs and N SUs, minimum SINRs $\overline{\gamma}_i$, $\forall i$, rate set $\mathbf{R}$ for SUs, p_m^{th}, p_m^{sup}, β, λ_T, and the desired tolerance ϵ.
2. Compute CGR using (2.4) for each SU link.
3. Check **IF** pre-admission criteria is met **THEN** go to next step.
4. OTHERWISE: Exit. The SU link interfere with PUs and is not admissible.
5. For each admissible SU link $i = 1$ to N

    (a) Compute instantaneous SINR γ_i.
    (b) **IF** Constraint (2.3) is satisfied **THEN** record the surplus $s_i(p_i)$.
    (c) OTHERWISE Adopt different rate $R_i \in \mathbf{R}$ or apply *post-admission* control process to reject the SU.

6. Continue this process until the desired accuracy is met, that is, $|p_i(t+1) - p_i(t)| \leq \epsilon$.

2.5 Numerical Results

To corroborate the theoretical analysis, cognitive radio enabled ad hoc network is considered where N SU links are established over randomly but uniformly distributed over an area of $1{,}000 \times 1{,}000$ m which coexist with $M = 1$ PU link. Initial input values are as primary bandwidth $W = 5\text{MHz}$, $p_i^{sup} = 1\text{W}$, interference limit $p_m^{th} = 0.2\text{W}$, random but feasible initial power p_i, noise power $n_i = 0.1\text{W}$, threshold $\lambda_T = 0.3$, and the tolerance $\epsilon = 0.0001$. SUs are assumed to be able to adapt to different modulation techniques such as BPSK, QPSK, 16-QAM, 64-QAM, 256-QAM and 1024-QAM, therefore $\mathbf{R} = \{1, 2, 4, 6, 8, 10\}$ bits/symbol. When there were $N = 11$ SU links active, variation of power allocation of SUs and their surplus are plotted in Figs. 2.2 and 2.3. The surplus for SU link is increasing with p_i when there is no penalty because of violation of interference limit.

Active SUs can adapt their transmit power and rates to maximize their surplus values while satisfying imposed QoS requirements for SUs and interference constraints for PUs. The numerical results obtained from simulations have shown that the algorithm converges to an Nash equilibrium. Surplus function was decreasing for some time for some users but that reaches to stable point after few iterations.

Fig. 2.2 Power variation for some SU links $N = 11$ SU links

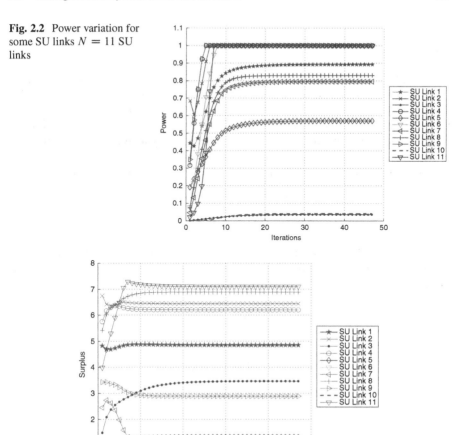

Fig. 2.3 Surplus variation for $N = 11$ SU links

2.6   Waiting Probability for DSA in TDMA CRNs

The "spectrum opportunity/hole" can be divided into different time slots so that many SUs can share the channel using time division multiple access (TDMA) technique [8]. When there are multiple SUs trying to access the same spectrum hole, they need to contend for the slots to get transmission opportunities. Note that, for a channel of bandwidth w_i Hz and receiver SINR γ_i, to successfully receive B bits of data, SU link needs either single slot with following time duration T or needs multiple slots that add up to the time T

$$T = \frac{B}{w_i \log_2(1 + \gamma_i)} \quad \text{s.} \tag{2.18}$$

As in legacy wireless networks, SUs become active to access idle bands using a Poisson process with rate $\lambda > 0$ and use the idle band with the service rate $\mu > 0$. Considering $S = \min\{S_1, S_2, \ldots, S_n\}$ as the shortest of the n service slot times, it is well known that the S is exponentially distributed with parameter $n\mu$ [6]. Now, it is of interest to find the probability that whether a newly active SU wait or the SU will be able to get one or more channel slots for transmission.

For an inter-arrival time X_i for SUs being active and considering $X_i = x$, we can write

$$Pr\{S \le x | X_i = x\} = 1 - e^{-n\mu x} \tag{2.19}$$

Let us consider $\overline{W}$ be an event that the next SU does not have to wait to access an idle slot, then the probability of getting an idle slot by next SU can be expressed as [8]

$$Pr[\overline{W}] = \frac{n\mu}{\lambda + n\mu} = 1 - \frac{\lambda}{\lambda + n\mu} \tag{2.20}$$

The waiting probability for next SU can be written as

$$Pr[W] = 1 - Pr[\overline{W}] = \frac{\lambda}{\lambda + n\mu} \tag{2.21}$$

Note that the waiting probability for SU depends on the rate of SUs being active, the number of available slots and the service rate.

To find the probability of getting one or more slots by a SU, Let us consider $S_{(1)} < S_{(2)} < \ldots < S_{(n)}$ as an ordered sequence of transmission completion time of SUs. Based on our assumption above, $S_{(1)} = \min\{S_1, S_2, \ldots, S_n\}$ and $S_{(2)} = \min\{S_2, S_3, \ldots, S_n\}$ are exponentially distributed with parameter $n\mu$ and $(n-1)\mu$ respectively. Consider $S_{(1)} = s$ and X be the inter-arrival time of the next SU, and assume that $X = s + x$ for some $x \ge 0$. Then, the probability that exactly one slot is available (i.e., the event E_1) for a given SU can be expressed as

$$Pr[\{E_1\}] = Pr[\{S_{(2)} > s + x\} | S_{(2)} > s]$$

Since the service time is memory-less, we can write as [8]

$$Pr[E_1] = Pr[\{S_{(2)} > x\}] = \frac{\lambda}{\lambda + (n-1)\mu} \tag{2.22}$$

Then, we find probability of having at least two channel slots in the case of more idle PU channels. In this case, SUs can choose the best channel that meets their rate requirement. Let us consider that E_0, E_1 and E_2 are the events that the SU finds 0,

1 or at least 2 idle slots respectively. Note that the events E_0, E_1 and E_2 are disjoint and

$$Pr[E_0 \cup E_1 \cup E_2] = 1 \tag{2.23}$$

Then we can write that [6]

$$\begin{aligned} Pr[E_2] &= 1 - Pr[E_0 \cup E_1] = 1 - (1 - Pr[\overline{E_0 \cup E_1}]) = Pr[\overline{E_0 \cup E_1}] \\ &= Pr[\overline{E_0} \cap \overline{E_1}] = Pr[\overline{E_0}]Pr[\overline{E_1}] = (1 - Pr[E_0])(1 - Pr[E_1]) \end{aligned} \tag{2.24}$$

We know that when SU has to wait for spectrum slots, the probability $Pr[E_0]$ is as (2.21), that is,

$$Pr[E_0] = \frac{\lambda}{\lambda + n\mu} \tag{2.25}$$

Then, from (2.24), (2.25) and (2.22), we can write the probability $Pr[E_2]$ as

$$Pr[E_2] = \frac{n\mu}{\lambda + n\mu} \times \frac{(n-1)\mu}{\lambda + (n-1)\mu} \tag{2.26}$$

2.6.1 Numerical Results

We plotted the variation of probability that the newly joining SU finds exactly one slot by using Eq. (2.22) or at least two idle slots by using Eq. (2.26) for a given arrival rate λ, service rate μ and for different number of (normalized) slots n as shown in Fig. 2.4. As expected, the probability of getting one or two slots increases with the number of slots in the systems. However, for a given number of slots in the system, the probability of getting two slots is less than that for one slot.

2.7 Summary

This chapter presented spectrum underlay spectrum sharing using code division multiple access technique by SUs where PUs are protected by interference limit and SUs are getting desired QoS. In the second part, probability of waiting for SUs for idle channel opportunities is presented for SUs that use time division multiple access technique for opportunistic spectrum access in CRN. Simulation results are used to evaluate the performance of the proposed approaches.

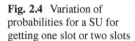

Fig. 2.4 Variation of probabilities for a SU for getting one slot or two slots

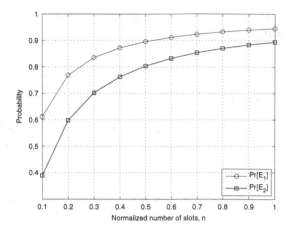

References

1. M. Andersin, Z. Rosberg, and J. Zander. Gradual Removals in Cellular PCS with Constrained Power Control and Noise. *Wireless Networks*, 2(1):27–43, 1996.
2. M. Chiang, P. Hande, T. Lan, and C. W. Tee. *Power Control Wireless Cellular Networks*. Foundations Trends Networking, 2008.
3. Hongyu Gu and Chenyang Yang. Power and Admission Control for UWB Cognitive Radio Networks. In *IEEE International Conference on Communications (ICC 2008)*, pages 4933–4937, Beijing, China, 2008.
4. J. Huang, R. Berry, and M. L. Honig. Distributed Interference Compensation for Wireless Networks. *IEEE Journal on Selected Areas in Communications*, 24(5):1074–1084, May 2006.
5. Long Bao Le and E. Hossain. Resource allocation for Spectrum Underlay in Cognitive Radio Networks. *IEEE Transactions on Wireless Communicationsg*, 7(12):5306–5315, Dec. 2008.
6. A. Papoulis and S.U. Pillai. *Probability, random variables and stochastic processes with errata sheet*. McGraw-Hill Science/Engineering/Math, 2001.
7. D. C. Popescu, O. Popescu, and D. B. Rawat. Gradient Descent Interference Avoidance with Target SIR Matching. In *Proceedings 5th Annual IEEE Consumer Communications and Networking Conference, CCNC'08*, Las Vegas, NV, Jan. 2008.
8. Danda B Rawat, Bhed B Bista, Gongjun Yan, and Sachin Shetty. Waiting probability analysis for opportunistic spectrum access. *International Journal of Adaptive and Innovative Systems*, 2(1):15–28, 2014.
9. Danda B Rawat, Bhed Bahadur Bista, and Gongjun Yan. Combined admission, power and rate control for cognitive radios in dynamic spectrum access ad-hoc networks. In *Network-Based Information Systems (NBiS), 2010 13th International Conference on*, pages 240–245, 2010.
10. D. M. Topkis. *Supermodularity and Complementarity*. Princeton University Press, 1998.

Chapter 3
Resource Allocation in Spectrum Overlay Cognitive Radio Networks

3.1 Introduction

Opportunistic spectrum access (OSA) is an emerging concept for spectrum overlay based spectrum sharing model where SUs identify spectrum opportunities in licensed bands and use them opportunistically without interfering with PUs [11, 19, 33, 35]. In spectrum overlay approach, as SUs are not allowed to co-exists with PUs in the same channel, they are required to either sense spectrum to find idle channels or search for idle channels in spectrum database [1, 7, 26]. Spectrum database can maintain geolocations of idle bands and provide a global view on entire frequencies which could be used to find best suitable channels for the SUs. Based on the global view of wideband RF regime, SUs could be granted a channel that has more adjacent channels so that SUs could implement channel bonding for higher data rates [12]. To prepare and update the spectrum database, database server can get spectrum occupancy information from PUs' infrastructure (e.g. base stations or access points) in a real-time basis. Alternatively, spectrum sensors (e.g., crowd sourcing for sensing [8, 37]) could be deployed to collect information about channel status. Based on the collected information, spectrum server can process data (with the help of cloud computing platform) create a spectrum maps for spectrum opportunities for different wireless networks such as satellite, WiMAX, Wi-Fi, cellular, TV, etc. Spectrum servers could be associated with a single or multiple spectrum servers (SSs) or brokers. When SU wants to access spectrum opportunities, it searches the geolocation database for idle bands. If there is an idle band available in given location and time, SU would access it. Otherwise, the SU has to wait until it finds spectrum opportunities that meets its needs.

In spectrum overlay approach, the SUs compete for better channels to meet their QoS requirements and the SSs compete to offer better services to many SUs and get more payoff. This process is formulated as a Stackelberg game [28]. This chapter presents a two-stage based Stackelberg game for OSA in CRN where SUs equipped with multiple transceivers dynamically access spectrum through SSs depending on

© The Author(s) 2015
D.B. Rawat et al., *Dynamic Spectrum Access for Wireless Networks*, SpringerBriefs in Electrical and Computer Engineering, DOI 10.1007/978-3-319-15299-8_3

their operating constraints. SUs are generally constrained by data rate subject to budget and QoS requirement while the SSs compete to provide competitive price for RF spectrum use by SUs subject to their spectral capacities. Furthermore, all SUs cannot be allowed to access idle bands for given time and location. Thus, we presented an analysis for admissibility check for SUs. Note that when SUs are equipped with multiple transceivers, they can have full-duplex communication when transmitting and receiving radios are tuned to non-overlapping channels. Similarly, when two different radios can be tuned for two different applications for example voice and data communications, wireless device can be used for both talk and surf simultaneously. When SSs have multiple channels and SUs have multiple radio for communications, they form multi-radio multi-channel (MRMC) wireless network [29]. The MRMC networks are extensively used in wireless mesh networks and broadcasting in wireless networks [10, 15, 30] to improve the throughput while avoiding interference. Numerical results are presented to evaluate the performance of the Stackelberg game.

There are several studies for spectrum overlay based opportunistic spectrum access in cognitive radio networks [4–6, 13, 16, 17, 21, 25, 31, 36]. None of these methods, take into account the impact of budget and QoS constraints of MRMC SUs in heterogeneous wireless environment where SUs maximize their achievable data rates and SSs maximize their revenues subject to their spectral capacity constraints.

3.2 Network Model and Problem Formulation

Networking model for OSA in CRN is shown in Fig. 3.1 where SUs and different wireless access technologies interact through a interface plane. This interface plane virtualizes different wireless networks such as TV, satellite, cellular and Wi-Fi networks and helps SUs to see all the channels as wide band wireless channels. SUs will have wireless connection and may not have any idea what networks they are using behind the interface plane. This interface can have multiple spectrum servers (SSs) who serve as a spectrum provider or broker. Chapter 4 presents cloud computing and storage for geolocation of idle bands.

The SSs could maintain an indexed table of geolocation database (x_a^b, y_a^b, z_a^b) with contour size r_a^b for a spectrum opportunity for ath position for bth network [7, 26, 27] as shown in Fig. 3.2. To avoid any harmful interference to PUs, the geolocation database with (x_a^b, y_a^b, z_a^b) is updated periodically in almost real-time. SSs could also estimate the time duration that the given band could be available for a given location. Note that when PUs resume their services while SUs are active, SSs switch the bands for SUs to other idle bands to protect PUs and not to drop the connection for SUs.

It is assumed that the SSs receive channel occupancy information from network infrastructures such as Wi-Fi access points, cellular base station, TV stations, etc. in almost real-time. Alternatively distributed spectrum sensors could sense the channels using sensing techniques as discussed in Chap. 1 and report the

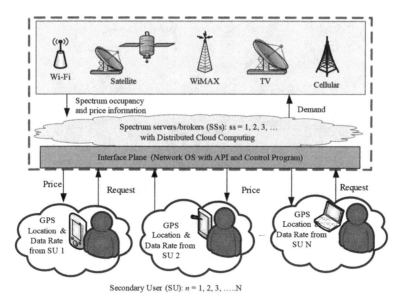

Fig. 3.1 Network model for network virtualization for opportunistic spectrum access in spectrum overlay approach

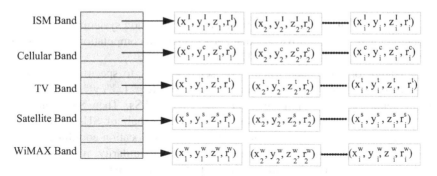

Fig. 3.2 Indexing of a geolocation in database with a radius of a contour for each network

information to the SSs. Based on the information SSs receive, they process data to find geolocation of idle bands and prepare a spectrum map as shown in Fig. 3.3 for different networks. Spectrum maps of spectrum opportunities are assumed to be updated in real-time with the help of distributed cloud computing [1, 27].

For spectrum opportunities in wide band RF spectrum regime, SUs may not be able to process real-time channel status reports as they are constrained by their limited memory, computational and power capacity. One alternative is to offload

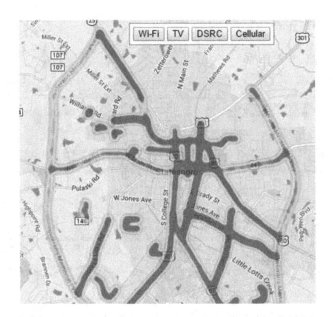

Fig. 3.3 Snapshot of Wi-Fi spectrum occupancy heat map on top of Google Map for streets of Statesboro, Georgia

the real-time channel status reports to external computation systems such as cloud computing platform that provides stream processing and computation due to its vast storage and computational capacity [26, 27].

To search for idle bands when a SU sends a query to spectrum database, it is assumed that it sends its geolocation information [e.g. $(x_i^{su}, y_i^{su}, z_i^{su})$] and QoS requirements (such as minimum data rate, etc.) to SSs. Then SSs search the geolocation database of spectrum opportunities on behalf of SUs to locate the idle bands for SUs by using similarity matching of geolocation coordinates of the idle bands and geolocation coordinates of SUs. The similarity matching of idle band location and SU's location could be performed by using either Euclidean distance (3.1) or cosine similarity (3.2). This similarity matching method works as an admission control method for SUs to protect PUs in CRN.

$$d_i^{su(b)} = \sqrt{(x_i^{su} - x_i^b)^2 + (y_i^{su} - y_i^b)^2 + (z_i^{su} - z_i^b)^2} \qquad (3.1)$$

When the distances $d_i^{su(b)}$ is less than the radius r_i^b or the specified tolerance, SUs could use the given band for given time and location. Thus, SSs will negotiate with the SU. If SU's requirements are met, it will access the given spectrum band for given time and location. SSs will allow a SU to use idle licensed bands who offers high revenue/benefit to them and less interference to other active users.

Alternatively, cosine similarity matching [14] can be used as an admission control for SUs by using geolocation database of idle bands $\mathbf{a}_b = (x_a^b, y_a^b, z_a^b)$ maintained by SSs and geolocation of SU $\mathbf{r}_s = (x_i^{su}, y_i^{su}, z_i^{su})$, i.e.,

$$M_{b,su} = cos(\theta) = \frac{\mathbf{a}_b . \mathbf{r}_s}{||\mathbf{a}_b|| \times ||\mathbf{r}_s||}$$

$$= \frac{x_a^b . x_i^{su} + y_a^b . y_i^{su} + z_a^b . z_i^{su}}{\sqrt{(x_i^{su})^2 + (y_i^{su})^2 + (z_i^{su})^2}\sqrt{(x_a^b)^2 + (y_a^b)^2 + (z_a^b)^2}}.$$

(3.2)

When both $\mathbf{a}_b$ and $\mathbf{r}_s$ represent the same geolocation, the value of $M_{b,su}$ is one implying that the user is a the center of the contour of the idle band. However, SUs cannot be at the center of the contour of the idle band all the time. We consider that SUs who satisfy $M_{b,su} \geq M_{b,su}^{th}$ criteria, they get permission to access the given spectrum band for given time and location. SUs who do not satisfy $M_{b,su} \geq M_{b,su}^{th}$ do not get admission to access the band. The $M_{b,su}^{th}$ is the threshold which can be adapted based on the sensing report. Note that when multiple SUs coexist in the same channel, they use code division multiple access type technology with suitable signatures to avoid interference to other active SUs [24, 34].

The main goal of this chapter is to present a solution to maximize SUs' achievable rates and the SSs' revenue while satisfying imposed constraints on both SUs and SSs using a two-stage Stackelberg game.

3.3 Two-Stage Stackelberg Game

In a two-stage Stackelberg game, the leaders (i.e., SSs) set the price using non-cooperative game and announce it to their followers (i.e., SUs). The followers choose their strategies to maximize their payoffs in terms of data rates. Note that the solution of Stackelberg game can be obtained by subgame perfect equilibrium (SPE) using backward induction [9, 18]. The analysis starts with SUs (Stackelberg followers) that they choose their strategies depending on the announced information by their leaders. Then analysis moves backward to SSs (Stackelberg leaders) where they set optimal pricing subject to their spectral capacity.

3.3.1 Follower Rate Maximization Sub-Game (FRMG)

The follower rate maximization game (FRMG) maximizes the data rate of SUs. In the two stage Stackelberg game, there are N SUs accessing K channels using K radios. There are $\mathcal{K} = \{1, 2, .., K\} \subseteq \{1, 2, .., K_c\}$ SSs where K_c represents the total number of channels available at SSs. It is assumed that SUs share the spectrum using code-division multiple access technology so that multiple SUs can coexist and transmit in the same channel simultaneously. Received signal at SU is assumed to be independent and identically distributed (i.i.d.) with zero mean and σ_s^2 variance. The noise that corrupts the received signal is assumed to be the AWGN and i.i.d. with variance σ_n^2. Then the SINR $\gamma_{n,k}$ of nth SU in kth channel can be expressed as

$$\gamma_{n,k} = \frac{\sigma_s^2}{\sigma_n^2}. \tag{3.3}$$

Note that when instantaneous SINR in (3.3) falls below required minimum SINR $\overline{\gamma}_{n,k}$, that is, $\gamma_{n,k} < \overline{\gamma}_{n,k}$, the given SU n in a given channel k will not have reliable communication or QoS requirement of SU will not be met. Feasible values of minimum SINRs can be chosen such that $\sum_{\forall n} \frac{\overline{\gamma}_{n,k}}{1+\overline{\gamma}_{n,k}} <$ processing gain of a given network k for code division multiple access based systems is satisfied [34].

For i.i.d. Raleigh fading channels probability density function of the instantaneous SINR $f_{\gamma_{n,k}}(\gamma_{n,k})$ is exponential [2, 3], then the probability of an outage can be expressed as [23]

$$P_{n,k}^{out} = Pr\{\gamma_{n,k} < \overline{\gamma}_{n,k}\} = 1 - \exp(-\frac{2^{R_{n,k}} - 1}{\hat{\gamma}_{n,k}}), \tag{3.4}$$

where $\hat{\gamma}_{n,k}$ is the average of instantaneous SINR values which is calculated as $\hat{\gamma}_{n,k} \geq -\frac{2^{R_{n,k}}-1}{\ln(1-\hat{P})}$ for the condition $P_{n,k}^{out} \leq \hat{P}$ with maximum allowed outage probability $\hat{P}$, and the $R_{n,k}$ is the data rate of nth SU in kth channel of bandwidth $w_{n,k}$ which is expressed as [32]

$$R_{n,k} = w_{n,k} \log_2(1 + \gamma_{n,k}) \tag{3.5}$$

Note that when a SU has $\gamma_{n,k} < \overline{\gamma}_{n,k}$ for a given observation period, SU should stop using the given channel. This protects other active SUs from any harmful interference created by the given SU who does not satisfy its requirement and still transmits with high power.

When a SU uses multiple channels through multiple radios, the sum-rate (which is regarded as a payoff for the follower/SU of the game) of the given SU n in all channels can be expressed as

$$u_{su,n} = \sum_{\forall k} R_{n,k} = \sum_{\forall k} w_{n,k} \log_2(1 + \gamma_{n,k}) \tag{3.6}$$

The sum rate in (3.6) is regarded as a payoff for the SU in FRMG game. Then, the three components of the FRMG game for payoff maximization are as

- Players: Active SUs are regarded as players of the game FRMG, i.e., $\mathcal{N} = \{1, 2, \dots, N\}$;

- Strategies: Strategies of the player are the actions taken by the SU to maximize its payoffs; and
- Payoff: Sum rate in (3.6) is the payoff of a player which is to be optimized based on the chosen strategies.

In the two-stage Stackelberg game, SSs, set the price for spectrum usage. Let us assume that c_k is the unit price set by a SS for the channel use k, and B_n denotes the budget of SU n. Note that the data rate and SINR are related by the expression $\gamma_{n,k} = 2^{R_k/w_{n,k}} - 1$ using Shannon–Hartley theorem for a given spectral efficiency of R_k [bit/s/Hz], thus the unit cost per-unit data rate and unit cost per-unit SINR can be used interchangeably. Furthermore, the budget B_n of the SU is related/similar to a payment for a data plan that wireless users pay to their cellular service providers, and it depends on transmit power of a SU n and its payment Pay for the wireless service [22, 38]. When SSs announce their price information to SUs, all SUs have information about a given set of prices $\{c_1, c_2, \ldots, c_K\}$. Then SU $n \in \mathcal{N}$ selects a channel provided by SSs that provides maximum payoff. This payoff maximization problem for a given SU could be formulated as

$$\underset{\gamma_{n,k}, \forall k \in \mathcal{K}}{\text{maximize}} \quad u_{su,n} = \underset{\gamma_{n,k}, \forall k \in \mathcal{K}}{\text{maximize}} \sum_{\forall k} w_{n,k} \log_2(1 + \gamma_{n,k})$$

$$\text{subject to} \quad \sum_{k=1}^{K} c_k \gamma_{n,k} \leq B_n; \tag{3.7}$$

$$\gamma_{n,k} \geq \overline{\gamma}_{n,k}; \quad \forall k \in \mathcal{K}.$$

$$p_{n,k} \leq \overline{p}_{n,k}; \quad \forall n \in \mathcal{N}, \forall k \in \mathcal{K}.$$

where $\overline{p}_{n,k}$ is upper limit of allowed power for nth SU in kth channel. The optimal power that a given SU needs to meet QoS requirement can be computed using (3.3) as $p_{n,k} = \overline{\gamma}_{n,k} \sigma_n^2$. Thus, to get maximum payoff, a given SU could choose the maximum allowed power as $p_{n,k} = min\{\overline{p}_{n,k}, \overline{\gamma}_{n,k} \sigma_n^2\}$. In this case, the power constraint $p_{n,k} \leq \overline{p}_{n,k}$ in problem (3.7) could be relaxed and the problem (3.7) can be expresses as

$$\underset{\gamma_{n,k}, \forall k \in \mathcal{K}}{\text{maximize}} \quad \sum_{\forall k} w_{n,k} \log_2(1 + \gamma_{n,k})$$

$$\text{subject to} \quad \sum_{k=1}^{K} c_k \gamma_{n,k} \leq B_n; \tag{3.8}$$

$$\gamma_{n,k} \geq \overline{\gamma}_{n,k}; \quad \forall k \in \mathcal{K}.$$

Note that the optimization problem (3.8) is convex in $\gamma_{n,k}$, and thus the solution of the problem has unique and optimal value.

The optimal SINR value for a given SU n in a channel k that maximizes the payoff in (3.8) is [28]

$$\gamma_{n,k} = \frac{\sum_{k=1}^{K} c_k + B_n}{K c_k} - 1. \tag{3.9}$$

However, to satisfy the data rate requirement of the SU imposed by SINR constraint, the optimal SINR should be

$$\gamma_{n,k} = max \left\{ \frac{\sum_{k=1}^{K} c_k + B_n}{K c_k} - 1, \overline{\gamma}_{n,k} \right\} \tag{3.10}$$

For nth SU with a radio in kth channel, a necessary condition to satisfy QoS imposed by $\gamma_{n,k} \geq \overline{\gamma}_{n,k}$, $\forall n$, $\forall k$ is [28]

$$B_n + \sum_{i=1,i \neq k}^{K} c_i \geq ((\overline{\gamma}_{n,k} + 1)K - 1)c_k, \qquad \forall n, \forall k \tag{3.11}$$

Conversely, user n needs $\gamma_{n,k} \geq \overline{\gamma}_{n,k}$ in channel k if

$$c_k \leq \frac{B_n + \sum_{i=1,i \neq k}^{K} c_i}{((\overline{\gamma}_{n,k} + 1)K - 1)}, \qquad \forall n, \forall k \tag{3.12}$$

Then, the game moves to leaders' subgame based on the SU's response in (3.9) where leaders set their optimal pricing parameters.

3.3.2 The Leader Price Selection Sub-Game (LPSG)

The SSs of the Stackelberg game are Leaders and they choose their price in a competitive manner without any cooperation with other SSs. Thus the LPSG is formulated as a non-cooperative game. In the LPSG, each SS is interested only in maximizing its outcome without paying attention to others. As in other game, a non-cooperative game consists of a set of players (i.e., SSs), a set of strategies/actions associated with each player, and an outcome associated with each player[20]. The actions of each player is to choose the best channel for a given SU by offering competitive price based on its own spectral capacity. Formally, the LPSG game for SSs is defined as

$$LPSG = \langle \overline{\mathcal{K}}, \{\mathcal{S}_k\}_{k \in \overline{\mathcal{K}}}, \mathcal{U}_{ss}(.) \rangle \tag{3.13}$$

where $\overline{K} = \{1, 2, \ldots, \overline{K}\} \subseteq K = \{1, 2, \ldots, K\}$ represents the players who are active SSs with $\overline{K} \leq K$ channels; S_k represents the set of pricing strategy of a given SS in a channel k given by (3.14); and $\mathcal{U}_{ss}(.) : \{S_1 \times .. \times S_{\overline{K}}\}$ represents the payoff given in (3.15) that maps strategy spaces to positive real numbers.

In LPSG, each player starts with reasonable arbitrary unit price $c_k > 0$ for a band k and announces it to the SUs through interface plane as shown in Fig. 3.1. This unit price for each channel is updated by each SS iteratively. For the next iteration, the unit price for a channel k based on the demand from SUs is updated as

$$c_k(t + 1) = c_k(t) + \beta_k \left(\sum_{\forall n} \gamma_{n,k} - S_k \right), \quad \forall k \tag{3.14}$$

where $\beta_k << 1$ is the price adjustment parameter for a channel k of a given SS (its approximate value is estimated in Fig. 3.4 in Sect. 3.5), $\sum_{\forall n} \gamma_{n,k}$ is the SUs demands, and $S_k = N_k(2^{R_k/w_{n,k}} - 1)$ is the spectral capacity of a channel k in which N_k is the total number of users that could be supported by a given channel in an ideal case. For example, a typical Wi-Fi access point (AP) supports approximately $N_k = 30$ users without significantly degrading network performance.

The SSs in LPSG try to select optimal pricing parameters to attract more SU requests so that they could maximize their own revenues. Whenever SSs update their price, they announce to SUs through interface plane. This price update process in LPSG is iterative and it continues until the price values converge, i.e., $c_k(t+1) \approx c_k(t)$, $\forall k$. Then, total revenue (payoff) function of a SS can be defined as

$$u_{ss}(c_k, \mathbf{c_{-k}}) = \sum_{\forall k} u_{ss,k}(c_k, \mathbf{c_{-k}}) \tag{3.15}$$

where a revenue function of a SS in a channel k, $u_{ss,k}(c_k, \mathbf{c_{-k}})$, for a given channel k and spectral capacity S_k, is

$$u_{ss,k}(c_k, \mathbf{c_{-k}}) = c_k \sum_{\forall n \in \mathcal{N}} \gamma_{n,k} \tag{3.16}$$

To study the existence of a Stackelberg equilibrium and to find the best response strategies of players in the two-stage Stackelberg game, we state the following formal definitions from game theory in the context of our problem.

Definition 3.1 (Stackelberg Equilibrium for the LPSG) The price set $\{c_1, c_2, \ldots, c_{\overline{K}}\}$ is a Stackelberg equilibrium for the LPSG for every SS in $k \in \overline{K}$, iif we have that [28]

$$u'_{ss,k}(c'_k, \mathbf{c_{-k}}) \geq u_{ss,k}(c_k, \mathbf{c_{-k}}), \quad \forall c'_k \in S_k \tag{3.17}$$

where $\mathbf{c_{-k}} = \{c_1, \ldots, c_{k-1}, c_{k+1}, \ldots, c_{\overline{K}}\}$ is the price set by SSs for channels other than k. Note that Stackelberg equilibrium is a strategy profile that neither the SSs (leaders) nor the SUs (followers) have the incentive to deviate their strategies from the equilibrium point.

Definition 3.2 (The Best Response for the LPSG) The best response of a SS in k to other players' strategies is the set

$$B_k = \{c_k | u'_{ss,k}(c'_k, \mathbf{c_{-k}}) \geq u_{ss,k}(c_k, \mathbf{c_{-k}}), \quad \forall c'_k \in \mathcal{S}_k\}. \tag{3.18}$$

Then, the revenue optimization problem for a SS can be expressed as

$$\underset{c_k, \forall k \in \overline{K}}{\text{maximize}} \quad u_{ss}(c_k, \mathbf{c_{-k}}) \tag{3.19}$$

$$\text{subject to} \quad \sum_{\forall n} \gamma_{n,k} \leq S_k, \tag{3.20}$$

$$c_k > 0, \quad \forall k \in \overline{K}. \tag{3.21}$$

Note that the sum of maximum values of $u_{ss,k}(c_k, \mathbf{c_{-k}}), \forall k$ maximizes the $u_{ss}(c_k, \mathbf{c_{-k}})$ in (3.15). Thus, we express the revenue optimization problem (3.19) for a SS in channel k as

$$\underset{c_k, \forall k \in \overline{K}}{\text{maximize}} \quad u_{ss,k}(c_k, \mathbf{c_{-k}}) \tag{3.22}$$

$$\text{subject to} \quad \sum_{\forall n} \gamma_{n,k} \leq S_k, \tag{3.23}$$

$$c_k > 0, \quad \forall k \in \overline{K}. \tag{3.24}$$

Revenue is increasing function in c_k and thus there exists an unique optimal solution. Optimal price c_k that maximizes the revenue of SS in k in (3.22) is given by [28]

$$c_k = \frac{\sum_{j=1, j \neq k}^{\overline{K}} c_j + \sum_{n=1}^{N} B_n}{N(\overline{K} - 1) + \overline{K} S_k} \tag{3.25}$$

3.3.3   The Best Response for the Stackelberg Game

This section present the best response of the two-stage Stackelberg game for both SUs and SSs. To show that the price value in (3.25) maximizes the SUs' payoffs in FRMG and SSs' revenues in LPSG, and is the best selection of the given SS for the channel k [28], we present the following analysis. To show that the c_k in (3.25) is the best response of the game, let us consider that the price increases from c_k to $c'_k = c_k + \delta_k$ for small $\delta_k > 0$ and for given prices $\mathbf{c_{-k}}$ of the other SSs rationally using non-cooperative game. Let us assume that the Eq. (3.12) is satisfied by both old price c_k and new price c'_k. We have that $c'_k > c_k$ for $\delta_k > 0$, and $\gamma_{n,k} - \gamma'_{n,k} > 0$ from (3.26) since $\gamma_{n,k} \leq \gamma'_{n,k}$ as

$$\gamma_{n,k} - \gamma'_{n,k} = \frac{c'_k - c_k}{c_k c'_k} \left[\frac{\sum_{j=1, j\neq k}^{\overline{K}} c_j + B_n}{\overline{K}} \right] \tag{3.26}$$

This implies that there will be no increase in SU's payoff because of increase in c_k. Thus, computed c_k is optimal for FRMG game.

Then, we move to SSs' $LPSG$ game to compute the difference in the revenues due to change in price from c_k to c'_k as

$$u'_{ss,k}(c'_k, \mathbf{c_{-k}}) - u_{ss,k}(c_k, \mathbf{c_{-k}}) = c'_k \sum_{\forall n} \gamma'_{n,k} - c_k \sum_{\forall n} \gamma_{n,k} \tag{3.27}$$

From (3.9) with c'_k and c_k and (3.27), we get that

$$u'_{ss,k}(c'_k, \mathbf{c_{-k}}) - u_{ss,k}(c_k, \mathbf{c_{-k}}) = -\frac{(\overline{K} - 1)}{\overline{K}} \delta_k \tag{3.28}$$

For $\delta_k > 0$, we can see in (3.28) that $u'_{ss,k}(c'_k, \mathbf{c_{-k}}) < u_{ss,k}(c_k, \mathbf{c_{-k}})$, which contradicts (3.17). Furthermore, when $\delta_k < 0$ there will be decrease in price resulting in $c'_k < c_k$, and thus there will be decrease in revenue.

Thus, it is concluded that the price value in (3.25) is the best response of the two-stage Stackelberg game.

3.3.4 The Existence and Uniqueness of the Equilibrium

In a non-cooperative game LPSG, if there are no channels with SSs, there exist no game. When there are channels with SSs, the price strategies are finite, non-empty and concave subset of some Euclidean space. Thus, there exist a Nash equilibrium. Furthermore, the $u_{ss,k}(c_k, \mathbf{c_{-k}})$ in (3.16) is continuous in c_k. The first order derivative of (3.16) is $\frac{\partial u_{ss,k}}{\partial c_k} = \sum_{\forall n \in \mathcal{N}} \gamma_{n,k}$ and the second order derivative of $u_{ss,k}(.)$ w.r.t. c_k is $0 \ \forall k \in \overline{K}$. Therefore, there exists a Nash equilibrium for this game. Furthermore, the result in Sect. 3.3.3 shows that there exist only one positive solution for the price selection LPSG game. Therefore, the Nash equilibrium of the LPSG game is unique and optimal, and thus there exists unique optimal equilibrium of the Stackelberg game.

3.4 The Algorithm

This section presents a formal algorithm based on the analysis presented above as Algorithm 3.1. The Step 9 in Algorithm 3.1 checks whether the algorithm is within the tolerance to ensure that the Stackelberg game reaches at optimal Nash equilibrium.

3.5 Numerical Results

This section presents simulation scenarios and numerical results obtained from simulations to evaluate the performance of the proposed algorithm.

First, the value of price adjusting parameter β_k is estimated. We simulated different scenarios with different values of N and $K = \overline{K}$ with bandwidth $w_{n,k} = 1$, $\forall n, \forall k$. We considered different ratios as $N/K = \{1, 2, 3, 4, 5, 6, 7, 8\}$ which represents whether the system us overloaded or underloaded in code division multiple access systems. The ratio $N/K = 1$ represents equally loaded while the ratio $N/K = 8$ represents heavily overloaded systems. We generated feasible values for other parameters such as $\overline{\gamma}_{n,k}$, c_k, etc. $\forall n, \forall k$ randomly and ran the Algorithm 3.1 to see the effect of different values of price adjusting parameter β_k. We plotted the number of iterations versus the β_k as shown in Fig. 3.4. As expected, we observed that when β_k was smaller (bigger), the algorithm took more (less)

Algorithm 3.1 Adaptation of strategies of SUs in FRMG and SSs in LPSG

1: **Input:** Initial $c_k(t = 1)$, $\forall k$, $\overline{\gamma}_{n,k}$, $\forall n$, $\forall k$ and tolerance ϵ.
2: **Output:** optimal payoff values $u_{ss,k}$ for SSs and $u_{su,n}$ for SUs.
3: **for** iteration t=2, 3, 4, … **do**
4:     **for** n=1, 2, 3 …, N **do**
5:         Solve (3.7) and send demanded $\gamma_{n,k}$ to SSs.
6:         Calculate the user payoff $u_{su,n}$.
7:     **end for**
8:     For each k, update the price using (3.14) and calculate the $u_{ss,k}$.
9:     **if** $(c_k(t + 1) - c_k(t) \geq \epsilon)$ **then**
10:        Announce $c_k(t + 1)$ to SUs and go to Step 3.
11:    **else**
12:        The price is converged within tolerance ϵ. //An optimal Stackelberg equilibrium has been reached. STOP.
13:    **end if**
14: **end for**

Fig. 3.4 Average number of iterations for the game to convergence to Nash equilibrium for different N/K values and variable β_k

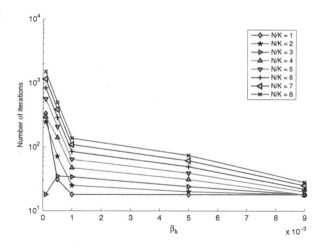

number of iterations. We noted that there is no significant difference when β_k decreases below 0.005 as shown in Fig. 3.4. The value of β_k can be any value that satisfies $\beta_k \ll 1$. However, in the following simulations we fixed $\beta_k = 0.0051$.

Then, the overloaded system with $N = 7$ SUs and $K = 4$ is considered where SUs use code division multiple access technology to avoid interference when multiple SUs coexists for opportunistic spectrum access. The minimum required SINRs values of SUs were generated as $\{\overline{\gamma}_{1,1} + \overline{\gamma}_{1,2} + \overline{\gamma}_{1,3} + \overline{\gamma}_{1,3}\} = 1.08$, $\{\overline{\gamma}_{2,1} + \overline{\gamma}_{2,2} + \overline{\gamma}_{2,3} + \overline{\gamma}_{2,4}\} = 2.16$, $\{\overline{\gamma}_{3,1} + \overline{\gamma}_{3,2} + \overline{\gamma}_{3,3} + \overline{\gamma}_{3,4}\} = 3.24$, $\{\overline{\gamma}_{4,1} + \overline{\gamma}_{4,2} + \overline{\gamma}_{4,3} + \overline{\gamma}_{4,4}\} = 4.33$, $\{\overline{\gamma}_{5,1} + \overline{\gamma}_{5,2} + \overline{\gamma}_{5,3} + \overline{\gamma}_{5,4}\} = 5.46$ $\{\overline{\gamma}_{6,1} + \overline{\gamma}_{6,2} + \overline{\gamma}_{6,3} + \overline{\gamma}_{6,4}\} = 6.51$, and $\{\overline{\gamma}_{7,1} + \overline{\gamma}_{7,2} + \overline{\gamma}_{7,3} + \overline{\gamma}_{7,4}\} = 7.50$, which are feasible for the considered wireless system. Initial price values for SSs were generated randomly and were $c_1 = 0.75$, $c_2 = 0.86$, $c_3 = 1.26$ and $c_4 = 1.76$. The specific values considered in the simulation for B_n were $\{B_1, B_2, B_3, B_4, B_5, B_6, B_7\} = \{1, 2, 3, 4, 5, 6, 7\}$ and spectral capacities of SSs were $S_1 = 60$, $S_2 = 62$, $S_3 = 65$ and $S_4 = 70$. We then plotted variation of SINRs and payoffs/utilities of SUs in Fig. 3.5a,b. SUs' SINRs are increasing as in Fig. 3.5a that give them higher payoffs or data rate as shown in Fig. 3.5b. All SINRs values for all SUs are higher than their minimum required SINRs.

We also plotted the variations of unit price and revenue of each SS in Fig. 3.6a,b. SSs use non-cooperative game while setting their prices and offer competitive price to SUs. Each SS reduced its price as shown in Fig. 3.6a using non-cooperative game while each SS maximized its revenue as shown in Fig. 3.6b by attracting more requests from SUs.

In Fig. 3.7, we plotted the SINR variation for each SU using Algorithm 1 and the algorithm in [16] by considering an identical simulation scenarios. The algorithm proposed in [16] maintains the minimum required SINR values for SUs, however, our algorithm gives higher SINRs for SUs than their corresponding minimum required values. Note that higher SINRs results in higher data rates for SUs.

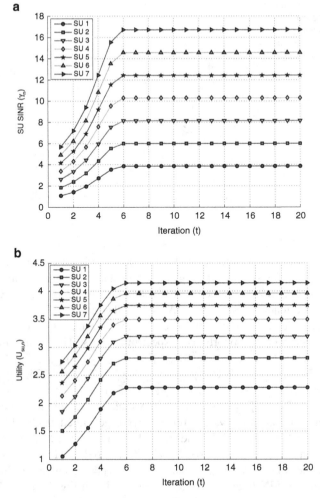

Fig. 3.5 Variation of SINRs and payoffs of SUs with $K = 3$ while algorithm reaches Nash equilibrium. (**a**) User SINR variation. (**b**) User payoff variation

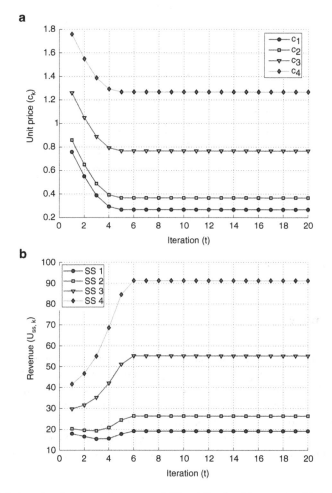

Fig. 3.6 Variation of unit price and revenue of SSs while algorithm reaches Nash equilibrium.
(**a**) Price variation. (**b**) Revenue variation

In conclusion, simulation results support that the proposed algorithm converges
to optimal and unique Stackelberg equilibrium that offer maximum payoffs (data
rates) to SUs and maximum revenues to SSs by satisfying their respective imposed
constraints. Note that the proposed approach also outperforms the existing related
methods [16, 28].

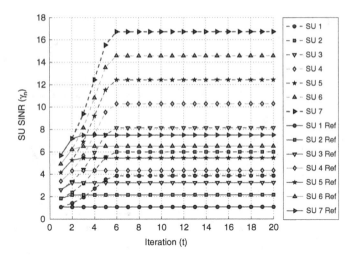

Fig. 3.7 Comparison of proposed approach with [16] for $\{\gamma_1, \gamma_2, \ldots, \gamma_6\} = \{1.08, 2.16, 3.24, 4.33, 5.46, 6.51, 7.5\}$

3.6 Summary

This chapter has presented a two stage Stackelberg game for opportunistic spectrum access by SUs equipped with multiple radios. The SUs' FRMG sub-game maximized SUs payoffs in terms of data rate subject to SUs' budget and QoS constraints. Meanwhile, the SSs LPSG sub-game maximized their revenues subject to their spectral capacities. This chapter has also provided the existence of an unique optimal point when the Stackelberg game is at equilibrium. The performance of the proposed approach is evaluated using numerical results obtained from simulations.

References

1. FCC, Second Memorandum Opinion and Order, ET Docket No FCC 10-174, September 2010.
2. Dan Avidor, Sayandev Mukherjee, Jonathan Ling, and Constantinos Papadias. On some properties of the proportional fair scheduling policy. In *Personal, Indoor and Mobile Radio Communications, 2004. PIMRC 2004. 15th IEEE International Symposium on*, volume 2, pages 853–858, 2004.
3. Jin-Ghoo Choi and Saewoong Bahk. Cell-throughput analysis of the proportional fair scheduler in the single-cell environment. *Vehicular Technology, IEEE Transactions on*, 56(2):766–778, 2007.
4. Yonghoon Choi, Hoon Kim, Sang-wook Han, and Youngnam Han. Joint resource allocation for parallel multi-radio access in heterogeneous wireless networks. *IEEE Transactions on Wireless Communications*, 9(11):3324–3329, 2010.
5. Lingjie Duan, Lin Gao, and Jianwei Huang. Cooperative spectrum sharing: a contract-based approach. *IEEE Transactions on Mobile Computing*, 13(1):174–187, 2014.

6. Basak Eraslan, Didem Gozupek, and Fatih Alagoz. An auction theory based algorithm for throughput maximizing scheduling in centralized cognitive radio networks. *IEEE Communications Letters*, 15(7):734–736, 2011.

7. O. Fatemieh, R. Chandra, and C. A. Gunter. Secure collaborative sensing for crowdsourcing spectrum data in white space networks. In *Proceedings DySPAN'10: IEEE International Dynamic Spectrum Access Networks Symposium*, April 2010.

8. Omid Fatemieh, Ranveer Chandra, and Carl A Gunter. Secure collaborative sensing for crowd sourcing spectrum data in white space networks. In *New Frontiers in Dynamic Spectrum, 2010 IEEE Symposium on*, pages 1–12, 2010.

9. Drew Fudenberg and David Levine. Subgame-perfect equilibria of finite-and infinite-horizon games. *Journal of Economic Theory*, 31(2):251–268, 1983.

10. Piyush Gupta and Panganmala R Kumar. The Capacity of Wireless Networks. *IEEE Transactions on Information Theory*, 46(2):388–404, 2000.

11. Zhu Ji and KJ Ray Liu. Cognitive radios for dynamic spectrum access-dynamic spectrum sharing: A game theoretical overview. *IEEE Communications Magazine*, 45(5):88–94, 2007.

12. Lei Jiao, Vicent Pla, and Frank Y Li. Analysis on channel bonding/aggregation for multi-channel cognitive radio networks. In *2010 European Wireless Conference (EW)*, pages 468–474, 2010.

13. Xin Kang, Rui Zhang, and Mehul Motani. Price-based resource allocation for spectrum-sharing femtocell networks: A stackelberg game approach. *IEEE Journal on Selected Areas in Communications*, 30(3):538–549, 2012.

14. Erwin Kreyszig, Herbert Kreyszig, and E J Norminton. *Advanced Engineering Mathematics*. Wiley, Hoboken, NJ, 2011.

15. Pradeep Kyasanur, Jungmin So, Chandrakanth Chereddi, and Nitin H Vaidya. Multichannel mesh networks: challenges and protocols. *IEEE Wireless Communications*, 13(2):30–36, 2006.

16. Tianming Li and Sudharman K Jayaweera. Analysis of linear receivers in a target sinr game for wireless cognitive networks. In *4th International Conference on Wireless Communications, Networking and Mobile Computing, 2008*, pages 1–5, 2008.

17. Shao-Yu Lien, Shin-Ming Cheng, Sung-Yin Shih, and Kwang-Cheng Chen. Radio resource management for qos guarantees in cyber-physical systems. *IEEE Transactions on Parallel and Distributed Systems*, 23(9):1752–1761, 2012.

18. Andreu Mas-Collel, Michael D Whinston, and Jerry Green. *Microeconomic Theory*. Oxford university press Oxford, 1995.

19. Jaime Lloret Mauri, Kayhan Zrar Ghafoor, Danda B. Rawat, and Javier Manuel Aguiar Perez. *Cognitive Networks: Applications and Deployments*. CRC Press, 2014.

20. John Nash. Non-cooperative games. *The Annals of Mathematics*, 54(2):286–295, 1951.

21. Dusit Niyato, Ekram Hossain, and Zhu Han. Dynamics of multiple-seller and multiple-buyer spectrum trading in cognitive radio networks: A game-theoretic modeling approach. *IEEE Transactions on Mobile Computing*, 8(8):1009–1022, 2009.

22. Miao Pan, Hao Yue, Chi Zhang, and Yuguang Fang. Path selection under budget constraints in multihop cognitive radio networks. *IEEE Transactions on Mobile Computing*, 12(6): 1133–1145, 2013.

23. A. Papoulis and S.U. Pillai. *Probability, random variables and stochastic processes with errata sheet*. McGraw-Hill Science/Engineering/Math, 2001.

24. Dimitrie C Popescu, Danda B Rawat, Otilia Popescu, and Mohammad Saquib. Game-theoretic approach to joint transmitter adaptation and power control in wireless systems. *IEEE Transactions on Systems, Man, and Cybernetics, Part B: Cybernetics*, 40(3):675–682, 2010.

25. D. B. Rawat and DC Popescu. Precoder adaptation and power control for cognitive radios in dynamic spectrum access environments. *IET Communications Journal*, 6(8):836–844, 2012.

26. Danda B Rawat, Sachin Shetty, and Khurram Raza. Secure radio resource management in cloud computing based cognitive radio networks. In *Parallel Processing Workshops (ICPPW), 2012 41st International Conference on*, pages 288–295, 2012.

27. Danda B. Rawat, Sachin Shetty, and Khurram Raza. Geolocation-aware Resource Management in Cloud Computing Based Cognitive Radio Networks. *International Journal of Cloud Computing (IJCC)*, 3(3), 2014. http://www.inderscience.com/offer.php?id=64765

28. Danda B Rawat, Sachin Shetty, and C. Xin. Stackelberg game based dynamic spectrum access in heterogeneous wireless systems. *IEEE Systems Journal*, 2014.

29. Jason Redi and Ram Ramanathan. The DARPA WNaN Network Architecture. In *Military Communications Conference 2012 – MILCOM 2011*, pages 2258–2263, 2011.

30. Wei Ren, Qing Zhao, Ram Ramanathan, Jianhang Gao, Ananthram Swami, Amotz Bar-Noy, M Johnson, and Prithwish Basu. Broadcasting in multi-radio multi-channel wireless networks using simplicial complexes. In *2011 IEEE 8th International Conference on Mobile Adhoc and Sensor Systems (MASS)*, pages 660–665, 2011.

31. S Senthuran, A Anpalagan, and O Das. Throughput analysis of opportunistic access strategies in hybrid underlay-overlay cognitive radio networks. *Wireless Communications Wireless Communications*, 11(6):2024–2035, 2012.

32. Claude Elwood Shannon and Warren Weaver. A mathematical theory of communication, 1948.

33. Min Song, Chunsheng Xin, Yanxiao Zhao, and Xiuzhen Cheng. Dynamic spectrum access: from cognitive radio to network radio. *IEEE Wireless Communications*, 19(1):23–29, 2012.

34. Pramod Viswanath, Venkat Anantharam, and David N. C. Tse. Optimal sequences, power control, and user capacity of synchronous cdma systems with linear mmse multiuser receivers. *IEEE Transactions on Information Theory*, 45(6):1968–1983, 1999.

35. Chunsheng Xin, Min Song, Liangping Ma, and Chien-Chung Shen. Performance analysis of a control-free dynamic spectrum access scheme. *IEEE Transactions on Wireless Communications*, 10(12):4316–4323, 2011.

36. Dan Xu, Eric Jung, and Xin Liu. Efficient and fair bandwidth allocation in multichannel cognitive radio networks. *IEEE Transactions on Mobile Computing*, 11(8):1372–1385, 2012.

37. Rui Zhang, Jinxue Zhang, Yanchao Zhang, and Chi Zhang. Secure crowdsourcing-based cooperative pectrum sensing. In *INFOCOM, 2013 Proceedings IEEE*, pages 2526–2534, 2013.

38. Liang Zheng, Chee Wei Tan, et al. Cognitive radio network duality and algorithms for utility maximization. *IEEE Journal on Selected Areas in Communications*, 31(3):500–513, 2013.

Chapter 4
Cloud-Integrated Geolocation-Aware Dynamic Spectrum Access

4.1 Introduction

This chapter presents cloud integrated dynamic spectrum access in cognitive radio networks where most of the computing and storing function are performed using data offloading to cloud computing platform. The SUs are considerably constrained by their limited power, memory and computational capacity when they have to make decision about spectrum sensing for wide RF band regime and dynamic spectrum access. The SUs in CRN have the potential to mitigate these constraints by leveraging the vast storage and computational capacity of cloud computing platform [19, 21]. Specifically, cloud computing based dynamic spectrum access has following advantages: (a) *Power saving in mobile devices*: As SUs search the spectrum opportunities in the geolocation database, power needed for SUs to sense the RF spectrum for wide range of bands will be saved. By leveraging the cloud computing and storage resources, SU mobile devices can extend their battery lifetime; (b) *No harmful interference to PUs*: Chances of mis-detection of spectrum opportunities can be significantly reduced when SUs are required to search the database instead of sensing and identifying spectrum opportunities by themselves. Furthermore, the aggressive SUs can be monitored and possibly penalized by incorporating a cloud assisted manager to oversee the overall system; (c) *Compliance with the requirements of the regulatory body*: Recently the FCC in the U.S. [1, 12] mandates that the SUs must search geolocation database for spectrum bands instead of sensing and identifying the spectrum opportunities themselves. Thus, the proposed approach follows the recent proposal by FCC and can be implemented easily in real systems; and (d) *Outsource computing on mobile devices*: Typical mobile devices used by SUs have limited computing capabilities that limits the scalability of cognitive networks. Using proposed approach, the computation performance of SUs is significantly enhanced by outsourcing the streaming computation tasks to the cloud computing systems.

© The Author(s) 2015   43
D.B. Rawat et al., *Dynamic Spectrum Access for Wireless Networks*, SpringerBriefs
in Electrical and Computer Engineering, DOI 10.1007/978-3-319-15299-8_4

To provide real-time status of spectrum bands and services, cloud computing environment is used due to its distributed computing and storage capabilities. Computation in the cloud can be carried out on Hadoops' MapReduce programming model [23] by using batch processing of vast amounts of data. However, in the CRN environment, to update spectrum occupancy status in real-time, recently, Hadoop Online Prototype [11], S4 [18], Hstreaming [2], Esper [5], and Storm [3] have emerged as alternatives to Hadoops' MapReduce programming model. The Storm model is preferred to cloud computing based CRN due to its real-time computational capabilities and it is open source tool.

It is noted that the related work assumes either SUs themselves are capable of sensing wide range of spectrum bands or high capacity single server dedicated to each frequency provides the spectrum occupancy information [14–16, 22, 24]. Former assumption may not be a suitable choice since sensing of wide frequency range for different wireless networks consumes most of the battery life of SU mobile devices. Latter assumption does not consider all frequency bands of each network. This might not be a good choice when SU wants to switch its transmission from one band to another to vacate the band for PUs.

This chapter focuses dynamic spectrum access in cloud-assisted dynamic spectrum access in CRNs [12, 19, 21] for both infrastructure-based and peer-to-peer communications [17, 25] for admissible SUs [19, 21].

4.2 System Model

The system model for cloud integrated dynamic spectrum access in CRN is shown in Fig. 4.1. We consider that the spectrum server uses cloud computing to process/analyze and store the geolocation (x_i^b, y_i^b, z_i^b) of an idle RF bands $b \in B \subseteq \{I(ISM), C(Cellular), T(TV), S(Satellite))\}$ in the geolocation database as in Fig. 4.1. The bands $I(ISM), C(Cellular), T(TV), S(Satellite)$ also contains several channel. For example, 2.4 GHz ISM has 14 total (11 usable) channels. Each geolocation (x_i^b, y_i^b, z_i^b) with a contour radius r_i^b [12] has an estimated time T_i^b which is the time that the band b is going to be idle for. To search spectrum opportunities in the geolocation database, each SU sends the request with its geolocation, payment P_i^s for the service and QoS demand in terms of data rate R_i^s to the spectrum server. When SUs are admitted based on their geolocation, they can have infrastructure based communications or peer-to-peer based communications as shown in lower left side of Fig. 4.1.

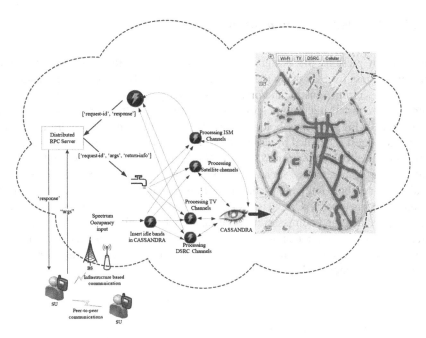

Fig. 4.1 Distributed RPC workflow, Storm topology for real time processing in cloud computing platform and SU communications

4.3 Computing Platform

For a wide band regime of heterogeneous wireless systems, spectrum occupancy information is huge and should be processed in a real-time basis. Each network can have a spectrum map as shown in right side of the Fig. 4.1. When servers located in the cloud receive spectrum occupancy information from sensors or PUs' infrastructures bolts of the Storm model process and store the data in a distributed manner.

The Storm system is used for processing spectrum data and SUs' requests in a real-time manner. Note that the Storm model is currently used by Twitter to process trending topics. The data in Storm is defined as an unbounded sequence of tuples. Spouts in Storm generates streams by reading them from an external sources. The generated streams are subscribed to one or more bolts which can produce additional streams. The bolts in Storm model are similar to the mappers and reducers of Hadoop but provide real-time data processing. The full set of spouts and bolts constitutes a topology. The topology needs to be executed to perform computation in Storm and it runs until you kill the process. The distributed RPC (DRPC) in Fig. 4.1 provides parallelism for computation. The DRPC server receives RPC requests, sends them to the Storm topology, Storm topology communicates with the Cassandra database and receives the response, and it sends the response back to the waiting client/SU.

4.3.1 Distributed Computation

This section presents the design of the Storm topology to implement dynamic spectrum access for SUs using their geolocations. Initially, the Cassandra database is empty. Spectrum sensors and/or PU infrastructure such as base station, access points, etc. report availability of spectrum opportunities to the 'spectrum occupancy input' bolt in Fig. 4.1. This bold takes input and update the Cassandra database. The SU's request is handled by a separate bolt that produces a output which becomes input for other bolts *Bolt for ISM band, Bolt for TV band, Bolt for DSRC, etc.* Note that each network has separate bolt to process the data for SUs and spectrum database. Each bolt implements the algorithm to process the data related to idle channels and geolocation of SUs. These bolts also emit idle channels that meets SUs' requirement and admissibility criteria. If there is only one idle channel, bolt returns the idle band as a response to SUs. If there are more than one idle channels, bolt identifies the idle channel that offers highest benefit to spectrum server and notifies that channel to SU [19–21].

4.3.2 Distributed Database

As described in the previous sections, cloud based spectrum servers receive the real-time status of spectrum bands from sensors or primary infrastructures. This information needs to be stored for searching by SUs in the storm topology. Based on the stored information, spectrum server could generate a spectrum map as shown in Fig. 4.1 for each network to visualize the status of RF spectrum of each network. This database state should be persistent in the presence of task failures which can be obtained by using a distributed database management systems. For distributed database in cloud computing environment, Riak [7], Cassandra [4] and Memcached [6] are possible candidates to store geolocation of idle channels. Apache Cassandra [4] is an open source and has proven robustness for real-time processing. It is commercially used by Twitter to process huge amount of data spread out across many servers while providing reliable services. Cassandra provides real-time processing by using its ability to store and access data in columns using distributed counters. Cassandra offers fault tolerance through replication of its data to multiple nodes and has capability of replacing failed node immediately with no downtime. In Cassandra, processing is decentralized and there is no single point of failure. Furthermore, Cassandra is less susceptible to network bottleneck problems and offers high availability, accessibility and accountability features needed for dynamic spectrum access in cognitive radio networks.

Storm model and Cassandra database are integrated by using available generic and configurable *backtype.storm.Bolt* implementation [8]. The *backtype.storm.Bolt* is dynamically configurable and it writes Storm Tuple objects to a Cassandra Column family in Cassandra database and the bolt writes the spectrum data to Cassandra cluster [19, 21].

4.4 Infrastructure-Based SU Communications

When SUs query for idle bands for a given location and time, the spectrum server applies admission control to SUs to protect PUs and offer better service to active SUs. In other words, each SU must go through admissibility check process [using Euclidean distance (3.1) or similarity matching (3.2)] to access idle bands as discussed in Chap. 3. Furthermore, servers can admit only few SUs who offer high benefit while fulfilling their own QoS requirements. This admission control ensures no harmful interference to PUs, maximum benefit to spectrum operators, and data rate satisfaction to SUs. The benefit maximization problem for spectrum operator could be formulated as [21]

$$\text{maximize} \quad \sum_{\forall s} B_i^{s(b)}$$

$$\text{subject to} \quad d_i^{s(b)} < (r_i^b + \epsilon), \tag{4.1}$$

$$\eta_i^b \leq R_i^s,$$

$$s_i^b \leq T_i^b,$$

where $B_i^{s(b)}$ is the benefit offered by SUs for a band b, $d_i^{s(b)}$ is the Euclidean distance between location of an idle frequency and location of SU, r_i^b is the radius of idle contour for a band b, ϵ is the tolerance, and η_i^b is a data rate threshold. The information about the geographic location, and estimated radius of the idle bands is stored in Cassandra database as shown in Fig. 4.1 where each bolt represent a hash like structure as shown in the Fig. 3.2. Furthermore, in order to better serve SUs and to protect PUs, the idle time T_i^b for a band b in a given geographic location could be estimated based on the spectrum usage history. The estimated time T_i^b is updated periodically based on spectrum occupancy information of a given band.

The effective benefit function is defined as the function of payment and interference [21], that is,

$$B_i^{s(e)} = \frac{P_i^s}{t_i^p + 1} \tag{4.2}$$

where P_i^s is the payment (that is charge for spectrum usage by SU) and

$$t_i^p = \begin{cases} 0, & \text{when } d_i^{s(b)} \leq r_i^b \\ C \times \left(d_i^{s(b)}\right)^n + r_i^b, & \text{otherwise} \end{cases} \tag{4.3}$$

is proportional to interference created to PUs where n is the exponent and its typical value is between 2 to 6 [13]. In (4.3), $0 < C < 1$ is the proportionality constant.

The value of t_i^p is zero when SU is within the idle spectrum contour. From (4.2) and (4.3), note that when interference t_i^p increases, the benefit $B_i^{s(e)}$ decreases and vice versa.

When multiple SUs try to access the same band, there will be a queue and this can be analyzed using $M/M/1$ queuing system where the SU request arrival rate to the spectrum server is λ_i and service rate is $\mu_i^{s(b)}$ for a given band b. The service time taken for a given SU s in band b is [10]

$$s_i^{s(b)} = \frac{1}{\mu_i^{s(b)} - \lambda_i} \tag{4.4}$$

Note that, the SU, if it has choices, will prefer to use a band that satisfy the following condition so that the band will be available until SU finishes its transmission.

$$T_i^b \geq s_i^{s(b)} \tag{4.5}$$

When a SU sends a request to access the spectrum opportunity, spectrum server can narrow down the search space by comparing the demanded data rate R_i^s of SU with the specified threshold η_i^b. Table 4.1 lists the typical maximum data rates offered by different wireless standards/networks.[1] For instance, when SU requests a data rate $R_i^s = 27$ Mbps, all networks that offer less data rate than 27 Mbps are excluded from the search space as those bands cannot fulfill the requested data rate requirement of the SU.

4.4.1 Numerical Results

To evaluate the performance of the proposed approach, simulation results are presented for underloaded, equally-loaded and overloaded scenarios. In underloaded case, the number of SUs is less than the number of idle channels and thus there will be less competition among SUs. In equally-loaded case, the number of SUs equal to the number of idle channels. Where as in overloaded case, the number of SUs is greater than the number of idle channels where more SUs compete (offering more payment) and contend for idle channels for transmission opportunities. When SU searches for idle bands, its admissibility based on geolocation is checked [e.g. using (3.1) or (3.2)]. If it is admissible, then the admissibility is checked based on SU's data rate requirement and effective benefit offered by SU to spectrum providers. In other words, spectrum servers admit the SUs whose geolocation are within the contour of idle channels and who offer reasonable benefit to servers while protecting PUs. The real-time distributed computing and storage is provided to

---

[1]It is also worth to mention that the data rate (speed) can vary based on the location of SU and modulation/transmission techniques implemented to transmit the information.

Table 4.1 Typical maximum data rates for different wireless standards/networks

Network standard	Max data rate (Mbps)
ZigBee	0.25
Bluetooth-v1.2	1
3G Cellular	2
Bluetooth-v2.0	3
TV-SDSL	3
802.11b Wi-Fi	11
TV-ADSL	12
Bluetooth-v3.0	24
Bluetooth-v4.0	24
TV-ADSL+2	26
802.11p DSRC	27
TV-VDSL	52
802.11a Wi-Fi	54
802.11g Wi-Fi	54
4G-LTE	100
4G-WiMAX	128
Satellite	155
802.11n Wi-Fi	600
4G-LTE Advanced	1,000
802.11ac Wi-Fi	1,300

dynamic spectrum access for SUs using Storm model and Cassandra database in the cloud computing environment. We considered a controlled simulation scenario with 40 SUs located randomly and contending for spectrum opportunities. To simulated all cases, we considered three different scenarios: (a) Underloaded scenario had 100 idle spectrum geolocations with 20 ISM, 34 Cellular, 26 TV and 30 Satellite geolocations; (b) Equally-loaded scenario had 40 idle spectrum geolocations with 10 ISM, 10 Cellular, 10 TV and 10 Satellite geolocations; and (c) Overloaded scenario had 20 idle spectrum geolocations with 5 ISM, 6 Cellular, 4 TV and 5 Satellite geolocations. Payment values offered by SUs are generated using uniform distribution which is similar to payment for data plans in traditional cellular phone systems.

We plotted the variation of benefit versus the number of active SUs as shown in Fig. 4.2 for overloaded, underloaded and equally loaded scenarios. As expected, total benefit increases in all cases with the increasing number of admitted SUs. Underloaded case offers highest benefit and overloaded case offers lowest benefit. In overloaded case, there will be high competition among SUs and only few SUs who offer high benefit get admission. Then spectrum providers set high threshold making few SUs admissible which results in lower benefit to spectrum provider as shown in Fig. 4.2.

Next, variation of admitted SUs in percentage among total number of contending SUs for idle channels is plotted in Fig. 4.3. Here, all SUs would not be admitted

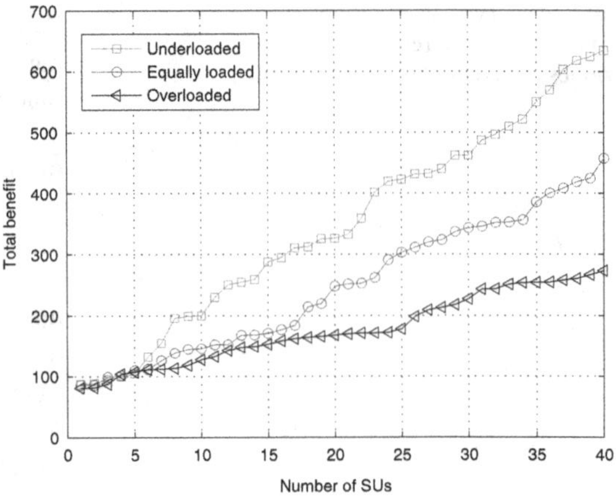

Fig. 4.2 The number of SUs versus the total benefit offered by admitted SUs

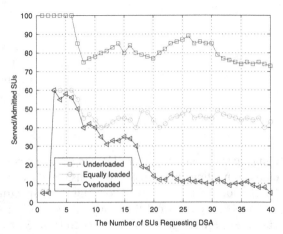

Fig. 4.3 The number of competing SUs versus the *admitted* SUs in percent

for channel access because of their geolocation mismatch with that of idle bands or high data rate demand for low benefit offer. The number of admitted SUs is lowest in overloaded scenario and highest in underloaded scenario among all three scenarios as shown in Fig. 4.3.

Figure 4.4 shows the variation of the number of admitted SUs versus the number of contending SUs for dynamic spectrum access. As expected, when the number of SUs increases, more number of SUs are admitted in all scenarios. Note that a given channel can handle only certain number of SUs at base station. Thus, the total number of admitted SUs must be checked against total allowed capacity of the channel.

Fig. 4.4 The number of SUs seeking for channel access vs. *admitted* SUs

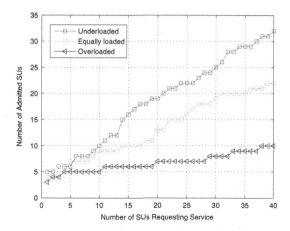

4.5 Distributed Power Adaptation Game (DPAG) for Peer-to-Peer SU Communications

In peer-to-peer SU communications, SUs first search geolocation database for idle bands and use suitable idle band for their communications as shown in Fig. 4.1. We start with overloaded system where number of SUs N is greater than the number of bands K (i.e., $N > K$) and many SUs compete for limited wireless channels/resources to get transmission opportunities. Analysis presented for overloaded could be easily used for equally loaded and underloaded systems. The SINR of the ith receiver[2] in kth channel of the band b is given as

$$\gamma_{i,k}^b = \frac{h_{i,i,k}^b}{\sum_{j \neq i} \alpha_{j,k} p_j^b h_{j,i,k}^b + n_0} \times p_i^b = G_{i,k}^b \times p_i^b \tag{4.6}$$

where $h_{i,j,k}^b$ is the path gain from transmitter of j-th pair to the receiver of ith pair via kth channel in the band b, $p_i^b \leq \overline{p_i^b}$[3] is the transmit power used by the i-th transmitter, and n_0 is the additive white Gaussian noise. In (4.6), $\alpha_{j,k} = 1$ when jth transmitter uses kth channel, and 0 otherwise.

Based on (4.6), the achievable data rate for the ith pair in the kth channel in the band b can be calculated as

$$R_{i,k}^b(\gamma_{i,k}^b(p_i^b)) = W_s \log_2(1 + \gamma_{i,k}^b) \tag{4.7}$$

where W_s is the bandwidth allocated to the SU i.

---

[2]In peer-to-peer based SU communications, we refer to SU i and SU link i interchangeably.

[3]The upper limit in transmission power is set by government authorities such as Federal Communications Commission (FCC) in the US.

For multi-user dynamic spectrum access, a Distributed Power Adaptation Game (DPAG) is expressed as

$$\mathcal{G} = \langle \mathcal{N}, \mathcal{S}, \{U_i^b(.)\}_{i \in N} \rangle$$

where

- $\mathcal{N} = \{1, 2, 3, \ldots, N\}$ represents a finite set of players who are the competing SUs (SU links) for spectrum opportunities,
- $\mathcal{S} = S_1 \times S_2 \times \ldots, S_j, \ldots, \times S_N$ defines the strategy profile space for active players/SUs. The strategy of a given SU is to choose a suitable transmit power for given channel k. The strategy profile of a player i can be represented as $s_i = [p_i^b, \mathbf{p}_{-i}^b]$ where $\mathbf{p}_{-i}^b$ is the strategy adapted by its opponents.
- $\{U_i^b(.)\}$ is a set utility function that the players/users want to maximize by choosing their suitable optimal strategies.

The utility function of a user i for the band b is given by

$$U_i^b(p_i^b) = [1 - P_{i,e}^b] R_{i,k}^b \tag{4.8}$$

where $P_{i,e}^b$ is a probability of error which can be computed as the probability of getting SINR below the desired SINR as

$$P_{i,e}^b = \text{Prob}\{\gamma_{i,k}^b < \gamma_i^{s*}\} \tag{4.9}$$

where γ_i^{s*} is the minimum desired SINR that is required to satisfy the quality-of-service of the SU i which can be calculated based on the minimum data rate R_i^s requested by SUs for a given bandwidth W_s as $\gamma_i^{s*} = (2^{R_i^s/W_s} - 1)$. That is, the successful/error-free transmission is defined by the condition

$$\gamma_{i,k}^b \geq \gamma_i^{s*} \tag{4.10}$$

To maximize the utility function, each SU iteratively chooses its best strategy given other players' strategies. The utility maximization problem can be written as

$$\underset{\gamma_{n,k}}{\text{maximize}} \quad U_i^b(p_i^b) \text{ in } (4.8)$$

$$\text{subject to} \quad Co1 : \gamma_{i,k}^b \geq \gamma_i^{s*}; \quad \forall s, \forall b, \tag{4.11}$$

$$Co2 : p_i^b \leq \overline{p_i^b}; \quad \forall s, \forall b.$$

The constraint $Co1$ ensures the quality-of-service requirements for a given active SU and the constraint $Co2$ makes sure that the given SU doesn't violate the average power limit in a given band.

To find the optimal power p_i^b by solving (4.11), we use a Lagrangian optimization approach with Lagrange's multipliers λ_1 and λ_2 as

$$L_{n,k} = (1 - P_{i,e}^b) \log_2(1 + \gamma_{i,k}^b) + \lambda_1(\gamma_{i,k}^b - \gamma_i^{s*}) - \lambda_2(p_i^b - \overline{p_i^b}) \tag{4.12}$$

After substituting (4.6) into (4.12), we compute first order optimality $\frac{\partial L_{n,k}}{\partial p_i^b} = 0$. For a given minimum SINR requirement γ_i^{s*}, we get

$$p_i^b = \frac{\gamma_i^{s*}}{G_{i,k}^b} \qquad (4.13)$$

But in order to comply with the FCC's power regulation, optimal power which is the best response of the game is given by

$$p_i^b = min\left(\overline{p_i^b}, \frac{\gamma_i^{s*}}{G_{i,k}^b}\right) \qquad (4.14)$$

where power of a given SU i in a band b depends on its desired minimum SINR γ_i^{s*}, effective channel gain $G_{i,k}^b$ and upper allowed limit on power $\overline{p_i^b}$.

Note that the power computed in (4.14) is the power at Nash equilibrium (NE) of the game if, for a SU game satisfies the following criteria [9]

$$U_i^b(p_1^b, p_2^b, \ldots, p_{j-1}^b, p_j^{b'}, p_{j+1}^b, \ldots p_N^b) \geq$$
$$\geq U_i^b(p_1^b, p_2^b, \ldots, p_{j-1}^b, p_j^b, p_{j+1}^b, \ldots p_N^b), \qquad (4.15)$$

where new power profile $p_j^{b'} \in S_j$. Note that NE is the most likely outcome of the game where no players get benefit by deviating their strategies and thus the NE is regarded as *socially optimal*. For distributed resource allocation games, *socially optimal* solution is obtained when the system satisfies (4.15).

Formally, algorithmic steps of the game, DPAG, are listed in the Algorithm 4.1.

Algorithm 4.1 Optimal power allocation for SUs

1: **Input**: initial transmit powers p_i^b, maximum power $\overline{p_i^b}$, minimum SINR value γ_i^{s*}.
2: **repeat**
3:     **for** Each SU pair i in a band b **do**
4:         Send query to search for idle bands. Spectrum servers check for admissibility for the SU link in b.
5:         **if** The SUs are admissible **then**
6:             SUs communicate using optimal power p_i^b in (4.14) and communicate.
7:             GOTO step 4 to periodically check the availability of spectrum.
8:         **else**
9:             SUs are not allowed to access the band b.
10:        **end if**
11:    **end for**
12: **until** All element of the set B are tested or B is NULL

4.5.1 Numerical Results

To evaluate the proposed approach, we consider a system where SUs become active randomly within a given geographic location and contend for opportunistic spectrum access by searching geolocation database of spectrum opportunities. When a SU searches for a suitable spectrum opportunity, spectrum server checks the admissibility conditions using (3.2) and (2.4) to allow or deny access for SUs.

We considered that six SUs for peer-to-peer communications are admitted for dynamic spectrum access where their required minimum SINRs of SUs are as $\gamma_i^{s*}=\{0.5, 0.45, 0.4, 0.35, 0.3, 0.2\}$ that represents different SUs with different QoS requirements. Each SU runs the Algorithm 4.1 to adapt its transmit power using DPAG game according to the estimated packet error rate based on its instantaneous SINR. SUs get optimal power allocation and meet the SINR requirement.

Once the system considered in this experiment reached the optimal Nash equilibrium, the sixth SU was dropped from the system leaving only five active SUs with their existing settings and continued running the algorithm to illustrate the tracking ability of the proposed the algorithm. When this new setup of five SUs reached at an optimal Nash equilibrium, a new SU who satisfies the admission control criteria is added to the system. The newly added SU had minimum required SINR of 0.25 which is different from previous SU's minimum SINR. After adding the new SU, the system had six SUs. Then, for the new setup with six SUs, we ran the algorithm.

Fig. 4.5 Power variation of SUs

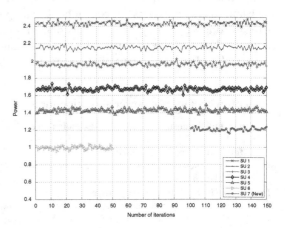

We plotted the power variation of all SUs as shown in Fig. 4.5. We noted that after 50th iteration sixth SU is dropped from the system and after 100th iteration new SU is added to the system as shown in Fig. 4.5. From 50th to 100th iteration there is no power variation for sixth SU as it is not active in the system.

Fig. 4.6 SINR variation of SUs

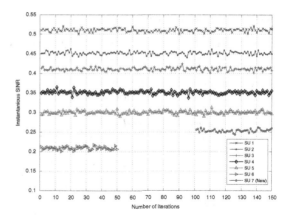

We have also plotted the SINR variation of each SU as shown in Fig. 4.6. We noted that the instantaneous SINR value of each SU is greater than its minimum required SINR which shows that the algorithm ensures the QoS requirement of each SU. As mentioned, there is no sixth user between 50th and 100th iteration and thus there is no SINR variation during that time. The newly added user added after 100th iteration has its instantaneous SINR value greater than its minimum required SINR 0.25 as shown in Fig. 4.6.

Note that the proposed algorithm can track the variable number of SUs with variable desired SINRs in the systems.

4.6 Summary

This chapter presented dynamic spectrum access for SUs with the help of cloud computing platform for processing huge data related to location of idle bands and location of SUs. Distributed processing and storage requirements and benefits that the cloud computing offers for dynamic spectrum access are presented. With the help of cloud computing, Storm model and Cassandra database, SUs are able to search and find idle bands for given time and location, and use those idle bands for either infrastructure-based communications or peer-to-peer based communications. Numerical results obtained from simulation are presented for performance evaluation of the proposed approaches.

References

1. FCC, Second Memorandum Opinion and Order, ET Docket No FCC 10–174, September 2010.
2. HStreaming Cloud: http://www.hstreaming.com/.
3. Storm: The hadoop of real-time processing: http://tech.backtype.com.

4. Apache Cassandra. http://cassandra.apache.org/, 2013. [Online; accessed 10-December-2013].
5. Esper. http://esper.codehaus.org/, 2013. [Online; accessed 10-December-2013].
6. Memcached. http://memcached.org/, 2013. [Online; accessed 10-December-2013].
7. Riak. http://basho.com/riak/, 2013. [Online; accessed 10-December-2013].
8. Storm cassandra integration. https://github.com/hmsonline/storm-cassandra, 2013. [Online; accessed 10-December-2013].
9. C. D. Aliprantis and S. K. Chakrabarti. *Games and Decision Making*. Oxford. University Press, New York, 2000.
10. D. Bertsekas and R. Gallager. *Data Networks*. Prentice Hall Inc., 1988.
11. Tyson Condie, Neil Conway, Peter Alvaro, Joseph M. Hellerstein, Khaled Elmeleegy, and Russell Sears. Mapreduce online. In *Proceedings of the 7th USENIX conference on Networked systems design and implementation*, NSDI'10, pages 21–21, Berkeley, CA, USA, 2010.
12. O. Fatemieh, R. Chandra, and C. A. Gunter. Secure collaborative sensing for crowdsourcing spectrum data in white space networks. In *Proceedings DySPAN'10: IEEE International Dynamic Spectrum Access Networks Symposium*, April 2010.
13. A. Goldsmith. *Wireless Communications*. Cambridge Univ Press, 2005.
14. Hiroshi Harada, Homare Murakami, Kentaro Ishizu, Stanislav Filin, Yoshia Saito, Ha Nguyen Tran, Goh Miyamoto, Mikio Hasegawa, Yoshitoshi Murata, and Shuzo Kato. A software defined cognitive radio system: cognitive wireless cloud. In *IEEE Global Telecommunications Conference, 2007. GLOBECOM'07*, pages 294–299, 2007.
15. D. T. Huang, Sau-Hsuan Wu, and Peng-Hua Wang. Cooperative spectrum sensing and locationing: a sparse bayesian learning approach. In *Global Telecommunications Conference (GLOBECOM 2010), 2010 IEEE*, pages 1–5. IEEE, 2010.
16. Chun-Hsien Ko, Din Hwa Huang, and Sau-Hsuan Wu. Cooperative spectrum sensing in tv white spaces: when cognitive radio meets cloud. In *Computer Communications Workshops (INFOCOM WKSHPS), 2011 IEEE Conference on*, pages 672–677, 2011.
17. Jinyang Li, Charles Blake, Douglas S.J. De Couto, Hu Imm Lee, and Robert Morris. Capacity of ad hoc wireless networks. In *Proc. of the 7th annual international conference on Mobile computing and networking*, MobiCom '01, pages 61–69, 2001.
18. Leonardo Neumeyer, Bruce Robbins, Anish Nair, and Anand Kesari. S4: Distributed Stream Computing Platform. In *Data Mining Workshops, International Conference on*, pages 170–177. IEEE Computer Society, 2010.
19. D. B. Rawat, S. Shetty, and K. Naqvi. Secure Radio Resource Management in Cloud Computing Based Cognitive Radio Networks. In *Proc. of the 41st International Conference on Parallel Processing (ICPP 2012)*, Pittsburgh, PA, September 12 2012.
20. Danda B Rawat, Sachin Shetty, and Khurram Raza. Game theoretic dynamic spectrum access in cloud-based cognitive radio networks. In *2014 IEEE International Conference on Cloud Engineering (IC2E 2014)*, pages 586–591, 2014.
21. Danda B Rawat, Sachin Shetty, and Khurram Raza. Geolocation-aware resource management in cloud computing-based cognitive radio networks. *International Journal of Cloud Computing*, 3(3):267–287, 2014.
22. Shie-Yuan Wang, Po-Fan Wang, and Pi-Yang Chen. Optimizing the cloud platform performance for supporting large-scale cognitive radio networks. In *Wireless Communications and Networking Conference (WCNC), 2012 IEEE*, pages 3255–3260, 2012.
23. T. White. *Hadoop: The Definitive Guide*. Yahoo Press, 2010.
24. Sau-Hsuan Wu, Hsi-Lu Chao, Chun-Hsien Ko, Shang-Ru Mo, Chung-Ting Jiang, Tzung-Lin Li, Chung-Chieh Cheng, and Chiau-Feng Liang. A cloud model and concept prototype for cognitive radio networks. *IEEE Wireless Communications*, 19(4):49–58, 2012.
25. Su Yi, Yong Pei, and Shivkumar Kalyanaraman. On the capacity improvement of ad hoc wireless networks using directional antennas. In *Proc. of the 4th ACM international symposium on Mobile ad hoc networking & computing*, MobiHoc '03, pages 108–116, 2003.

Chapter 5
Resource Allocation for Cognitive Radio Enabled Vehicular Network Users

5.1 Introduction

Vehicular communication networks are expected to provide safety and comfort services to passengers and drivers. To support diverse set of services and applications, vehicular network is expected to used variety of wireless access technologies for Vehicle-to-Vehicle (V2V) and Vehicle-to-roadside (V2R) communications. V2R communication could introduce high delay as roadside communication unit relays the information from source vehicle to destinations vehicles. Thus, for time critical messages, V2R based communication is not suitable to notify drivers in a timely manner. V2V based communication in VANET is performed in a peer-to-peer basis and the intended vehicles could exchange their information directly using single hop or multi-hop communications. In this case, performance of VANET depends on connectivity among vehicles since reliable connectivity for single hop or multi-hop communication is very important to forward time-critical information. The connectivity in VANET is directly related to density of vehicles, relative speed of the vehicles, association time of wireless technology, and transmission range and frequency bands used by vehicles. Time duration (T_D) for association (sensing/searching and connection setup) and information exchange depends on transmission range (R) and relative speed of vehicles (V_r) as

$$T_D = \frac{R}{V_r} \tag{5.1}$$

For a given transmission range, higher relative speed results in lower time duration for overlap in communication range of vehicles and vice versa.

Recent related work concerning the network connectivity in VANET includes [3, 4, 8, 9, 19, 20, 25–29]. None of the work considers the frequency agile vehicular communications. Furthermore, dynamic spectrum access for VANET

© The Author(s) 2015
D.B. Rawat et al., *Dynamic Spectrum Access for Wireless Networks*, SpringerBriefs in Electrical and Computer Engineering, DOI 10.1007/978-3-319-15299-8_5

users (aka SUs) has been proposed in [5, 7, 12, 22]. None of the methods considers transmission range adaptation and the effect of sensing and association time for V2V communications. Furthermore, none of the work in the literature considers connectivity for both one-way traffic flow and two-way traffic flow for spectrum-agile systems (where, for instance, vehicles switch channels between 5 GHz/2.4 GHz ISM band to 5.9 GHz DSRC band and vice versa). There is an IEEE 802.11p standard for vehicular communications which works well for dedicated short range communications (up to 1,000 m) in 5.9 GHz bands. However, seven channels (including one control channel) in 802.11p standard are not sufficient to support different VANET services and applications.

This chapter presents an analysis for V2V connectivity for vehicles traveling in opposite directions as well as in same direction. This chapter also presents software defined radio based dynamic spectrum access for vehicle-to-vehicle communication where we assume that transceivers mounted in each vehicle are capable of switching channels and adapting other transmit parameters (transmit power, modulation, etc.). As mandated by FCC, we assume that the geolocation of idle bands are stored in a database [1, 17, 18] as discussed in Chap. 4. Each vehicle uses a given destination to find the best route and uses this route information to search the spectrum database for the best idle channel. When two vehicles are within the communication range, they setup a communication link and exchange their data with the help of software defined radios. Numerical results obtained from simulations are presented to support mathematical analysis.

5.2 Networks Model

Each vehicle is assumed to be equipped with electronic devices based on the U.S. National Highway Transportation Safety Administration (NHTSA) ruling [2] to participate in communications. It is also assumed that these electronics devices are software defined and capable of switching channels and also capable of adapting transmit and receive parameters. Individual vehicles who participate in vehicular communications are required periodically broadcast their status information such as location, speed, direction, etc. in the network [24].

A typical network model is given in Fig. 5.1 where each vehicle is equipped with two radios. One radio (R1) is used for searching spectrum opportunities in a database and the other radio (R2) is used for exchanging data in vehicular network. Radio R1 is assumed to be always connected to internet and capable of searching spectrum database and working as a GPS through an app. For instance, for a route from location A to B in Fig. 5.1, channels , 6 and 1 are used by residential Wi-Fi users (aka PUs) throughout the first half of the route and thus vehicular users (SUs) are not allowed to use those channels. However, SUs can use any other channels than the channels used by PUs.

To process huge amount of data for different channels, we leverage cloud computing and cloud storage as discussed in Chap. 4. The radio R1 of each

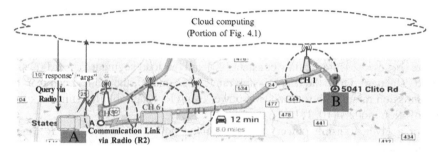

Fig. 5.1 System model for cloud-assisted GPS-driven dynamic spectrum access in cognitive vehicular network

vehicle queries the spectrum database to find spectrum opportunities and selects the best channel that meets its data rate requirements. This search process is periodic throughout the route to get updated information about idle channels and to protect PUs.

5.3   Analysis for Connectivity in VANET

To analyze the connectivity in VANET based on number of vehicles, each vehicle estimates number of its neighbors N_e based on the received periodic broadcast information in VANET [14, 21].

Each vehicle can also estimate the number of vehicles that could be present on a given road segment of length L for given transmission range and given number of lanes as [14]

$$N_t = \frac{L}{S_d} N_{Ln} \tag{5.2}$$

where S_d is the safety separation distance between vehicles, N_{Ln} is total number of lanes on the road.

Using estimated number of neighboring vehicles N_e and total number of vehicles N_t that could be present within the given transmission range of a vehicle, each vehicle can estimate the normalized vehicle density as

$$K_e = \frac{N_e}{N_t} \tag{5.3}$$

Then, based on the estimated normalized vehicle density, each vehicle can adjust its transmission range as [14, 21]

$$R = \min\{L(1 - K_e), \sqrt{\frac{L \ln L}{K_e}} + \alpha L\} \tag{5.4}$$

where $0 < \alpha < 1$ is traffic flow constant [23] and the road segment $L = 1,000$ m in DSRC standard. Once the transmission range is estimated, the vehicle can map the transmission range with transmission power using look up table for suitable signal prorogation models [10, 14].

For a given transmit power p_t, the received power p_r at distance R can be calculated as [10, 16]

$$p_r = p_t \underbrace{G_t G_r h_t h_r}_{G_e} \left(\frac{1}{4\pi}\right)^2 \frac{\lambda_w^2}{R^{\alpha_p}} = p_t G_e \frac{\lambda_w^2}{R^{\alpha_p}} \tag{5.5}$$

where h_t and h_r are respectively height of transmit and receive antenna, G_t and G_r are respectively transmit and receive antenna gains, λ_w is the wavelength (e.g., $\lambda_w = 5.08$ cm for 5.9 GHz DSRC band and $\lambda_w = 12.50$ cm for 2.4 GHz ISM band, etc.), and $\alpha_p \in [2, 4]$ is the path loss exponent. For a given transmission range in (5.4), the transmit power from (5.5) is expressed as [10]

$$p_t = \frac{p_r}{G_e \lambda_w^2} R^{\alpha_p} \tag{5.6}$$

From (5.6), we see that the transmit power level p_t is a function of frequency (or wavelength) that the vehicular users choose to communicate in spectrum-agile VANETs and transmission range. If vehicles would like to use same transmission range for different bands, they need to adapt their transmit power to cover the same range.

Using (5.4) and (5.1), time duration for given relative speed can be calculated. Note that the time duration computed in (5.1) is used for association (searching common channel and setting up a link) and message transmission.

Probability of successful information exchange, which depends on association time A and data rate of a given technology, relative speed of communicating vehicles, and size of the message to be transmitted, is given as [13, 16]

$$P_s = Pr\{A + B \leq T_D\} \tag{5.7}$$

where $B = S/D_r$ is the time needed to exchange the complete message of size S with data rate D_r using a single-hop communication and A is association time which includes sensing/searching time for idle bands along with connection setup time for dynamic spectrum access. If the condition $A + B \leq T_D$ is not satisfied, there will be failure in VANET communication. Note that exchange of partial information in VANET may not make any sense for users.

5.4 Numerical Results

In order to illustrate the performance of the analytical model presented in previous section, the numerical results are presented below.

In Fig. 5.2, the variation of total time duration versus the *relative* speed of vehicles for 1,000 m (upper limit in DSRC) is plotted. For vehicles traveling in opposite directions with a *relative* speed of $V_r = 141$ mph, for a single-hop communication, total time duration is $T_D = 1,596$ ms when transmission-range is $R = 1,000$ m. Note total time duration T_D increases when relative speed decreases and vice versa as in shown in Fig. 5.2.

Note that the time duration for vehicles traveling in same direction will have longer time depending on their relative speed and communication range. When there relative speed is zero, they remain within communication range of each other for virtually infinitely long time.

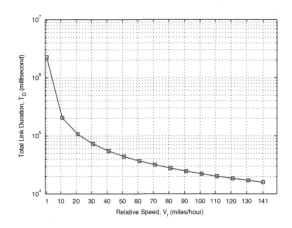

Fig. 5.2 Variation of link duration for different relative for speeds when DSRC transmission range $R = 1,000$ m

In Fig. 5.3, we plotted the time duration versus the relative speed for different association time values (that is the sum of spectrum search time and connection setup time). As expected, for a given relative speed, when association time increases, vehicles get lower time for actual communications as shown in Fig. 5.3. Note that the shorter time duration for communications results in smaller amount of message being exchanged between vehicles.

In Fig. 5.4, the variation of total message size that can be transferred after successful connection setup with 27 Mbps (max data rates in DSRC standard) for transmission range $R = 1,000$ m and $R = 25$ m. As expected, the message size is smaller when range is shorter since message size $= D_r \times (T_D - A)$

When multiple vehicles (SUs) share same channel for communications, the data rate per vehicular CR user in a given channel n for achievable data rate R_a (e.g. 11 Mbps in 802.11a, 27 Mbps in 802.11p DSRC, 54 Mbps 802.11g, etc.) can be computed as [11]

Fig. 5.3 Time duration vs.
the relative distance for
different association time
(search time for idle spectrum
and connection setup time)
for a given transmission range
of 1,000 m

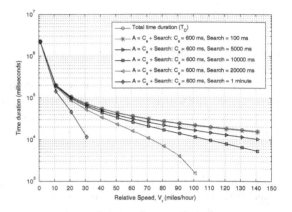

Fig. 5.4 Exchanged message
size after successful
association versus the
different relative speeds

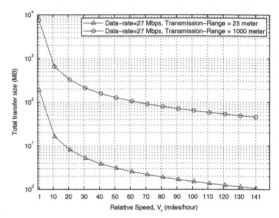

$$R_n^1 = \frac{R_a}{N_e} \cdot (1 - p_p) \tag{5.8}$$

where p_p is the probability of a channel being absent in the spectrum database (or a
PU being present in the given channel).

SU selects the best channel among available ones that offers highest data rate
and satisfies its minimum data rate (R_T) requirements. Furthermore, the radio R1
periodically queries the database to find the best channel that satisfies $R_n^1 \geq R_T$.

Along the line of [6, 15], the average time for successful data delivery can be
expressed in terms of expected transmission time (TT) for data size S bits for a data
rate R_n^1 in link n as

$$TT = \frac{1}{1 - p_f} \cdot \frac{S}{R_n^1} \tag{5.9}$$

where p_f is the probability of transmission failure in a link n.

Fig. 5.5 Variation of different metrics in cloud-assisted DSA in VANETs. (**a**) R_n^1 vs. N_e for $p_p = 0.5$. (**b**) TT vs. size S for $N_e = 50$

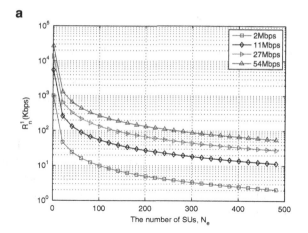

a

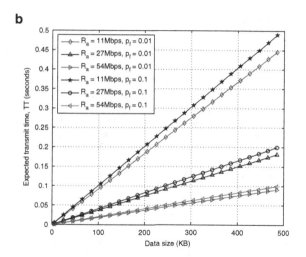

b

We plotted the variation of per user rate R_n^1 in a given channel n versus the number of VANET users (N_e) for different data rates (R_a) when $p_p = 0.5$ as shown in Fig. 5.5a. We observed that the per user rate R_n^1 decreases with number of users as in Fig. 5.5a. Next, we plotted the variation of expected transmission time (TT) versus the data size for $N_e = 50$, and different p_f and R_a values as shown in Fig. 5.5b. We observed that the expected transmission time increases when data size increases. Transmit time also increases when probability of failure increases as shown in Fig. 5.5b.

5.5   Summary

This chapter presented an analysis for VANET connectivity for vehicles traveling in opposite directions as well as for vehicles traveling in same direction. Analysis showed that the VANET connectivity depends on vehicle density, association time of the wireless technology, transmission range, relative speed of the vehicles and frequency bands used for communications. In a cloud assisted GPS-driven dynamic spectrum access in V2V communications, a wireless device mounted in a vehicle (SU) searches the spectrum database for idle channels and chooses the best channel to maximize its data rate. When two vehicles are within the communication range, they setup a communication link and exchange their data with the help of software defined radios. The performance depends on the association time, time for searching idle bands, and data size.

References

1. FCC, Second Memorandum Opinion and Order, ET Docket No FCC 10-174, September 2010.
2. The US NHTSA Final Regulatory Evaluation. http://www.nhtsa.gov/DOT/NHTSA/Rulemaking/Rules/AssociatedFiles/EDRFRIA.pdf.
3. Maen Artimy. Local Density Estimation and Dynamic Transmission-Range Assignment in Vehicular. *Intelligent Transportation Systems, IEEE Transactions on*, 8(3):400–412, 2007.
4. Neelakantan Pattathil Chandrasekharamenon and Babu AnchareV. Connectivity analysis of one-dimensional vehicular ad hoc networks in fading channels. *EURASIP Journal on Wireless Communications and Networking*, 2012(1):1–16, 2012.
5. Nan Cheng, Ning Zhang, Ning Lu, X Shen, J Mark, and Fuqiang Liu. Opportunistic spectrum access for CR-Vanets: A game theoretic approach. *IEEE Tran. Vehicular Technology*, 63(1), 2014.
6. Richard Draves, Jitendra Padhye, and Brian Zill. Comparison of routing metrics for static multi-hop wireless networks. In *ACM SIGCOMM Computer Communication Review*, volume 34, pages 133–144, 2004.
7. Gustavo Marfia, Marco Roccetti, Alessandro Amoroso, Mario Gerla, Giovanni Pau, and J-H Lim. Cognitive cars: constructing a cognitive playground for VANET research testbeds. In *Proceedings of the 4th International Conference on Cognitive Radio and Advanced Spectrum Management*, page 29, 2011.
8. Valery Naumov and Thomas R Gross. Connectivity-aware routing (CAR) in vehicular ad-hoc networks. In *INFOCOM 2007*, pages 1919–1927, 2007.
9. Sooksan Panichpapiboon and Wasan Pattara-Atikom. Connectivity requirements for self-organizing traffic information systems. *Vehicular Technology, IEEE Transactions on*, 57(6):3333–3340, 2008.
10. T.S. Rappaport. *Wireless Communications: Principles and Practice*. Prentice Hall PTR New Jersey, 2002.
11. D. B. Rawat, S. Reddy, N. Sharma, B. B. Bista, and S. Shetty. Cloud-assisted GPS-driven Dynamic Spectrum Access in Cognitive Radio Vehicular Networks for Transportation Cyber Physical Systems. In *IEEE Wireless Communications and Networking Conference (IEEE WCNC 2015)*, New Orleans, LA, USA, March 2015.
12. Danda B Rawat, Bhed B Bista, and Gongjun Yan. CoR-VANETs: Game Theoretic Approach for Channel and Rate Selection in Cognitive Radio VANETs. In *Broadband, Wireless Computing, Communication and Applications (BWCCA), 2012 Seventh International Conference on*, pages 94–99, 2012.

13. Danda B Rawat, Bhed B Bista, Gongjun Yan, and Stephan Olariu. Vehicle-to-vehicle connectivity and communication framework for vehicular ad-hoc networks. In *Complex, Intelligent and Software Intensive Systems (CISIS), 2014 Eighth International Conference on*, pages 44–49, 2014.

14. Danda B Rawat, Dimitrie C Popescu, Gongjun Yan, and Stephan Olariu. Enhancing VANET performance by joint adaptation of transmission power and contention window size. *Parallel and Distributed Systems, IEEE Transactions on*, 22(9):1528–1535, 2011.

15. Danda B. Rawat, Swetha Reddy, Nimish Sharma, and Sachin Shetty. Cloud-assisted Dynamic Spectrum Access for VANET in Transportation Cyber-Physical Systems. In *IEEE IPCCC 2014*, December 2014.

16. Danda B Rawat and Sachin Shetty. Enhancing connectivity for spectrum-agile vehicular ad hoc networks in fading channels. In *Intelligent Vehicles Symposium Proceedings, 2014 IEEE*, pages 957–962, 2014.

17. Danda B Rawat, Sachin Shetty, and Khurram Raza. Geolocation-aware resource management in cloud computing-based cognitive radio networks. *International Journal of Cloud Computing*, 3(3):267–287, 2014.

18. Danda B Rawat, Sachin Shetty, and C. Xin. Stackelberg game based dynamic spectrum access in heterogeneous wireless systems. *IEEE Systems Journal*, 2014.

19. Danda B. Rawat, G. Yan, B. Bista, and M. C. Weigle. Trust On the Security of Wireless Vehicular Ad-hoc Networking. *Ad Hoc & Sensor Wireless Networks (AHSWN) Journal*, 2014. in press.

20. Danda B Rawat and Gongjun Yan. Infrastructures in vehicular communications: Status, challenges and perspectives. In M .Watfa, editor, *Advances in Vehicular Ad-Hoc Networks: Developments and Challenges*. IGI Global, USA, 2010.

21. Danda B Rawat, Gongjun Yan, Dimitrie C Popescu, Michele C Weigle, and Stephan Olariu. Dynamic adaptation of joint transmission power and contention window in vanet. In *Vehicular Technology Conference Fall (VTC 2009-Fall), 2009 IEEE 70th*, pages 1–5. IEEE, 2009.

22. Danda B Rawat, Yanxiao Zhao, Gongjun Yan, and Min Song. CRAVE: Cognitive Radio Enabled Vehicular Communications in Heterogeneous Networks. In *IEEE RWS'2013*, pages 190–192, January 2013.

23. Roger Roess, Elena Prassas, and William McShane. *Traffic Engineering*. 2010.

24. Raja Sengupta and Qing Xu. DSRC for Safety Systems. volume 10, pages 2–5. California PATH – Partners for Advanced Transit and Highways, 2004.

25. Sok-Ian Sou and Ozan K Tonguz. Enhancing VANET connectivity through roadside units on highways. *Vehicular Technology, IEEE Transactions on*, 60(8):3586–3602, 2011.

26. Gongjun Yan and Stephan Olariu. A probabilistic analysis of link duration in vehicular ad hoc networks. *Intelligent Transportation Systems, IEEE Transactions on*, 12(4):1227–1236, 2011.

27. Saleh Yousefi, E Altmaiv, Rachid El-Azouzi, and Mahmood Fathy. Connectivity in vehicular ad hoc networks in presence wireless mobile base-stations. In *7th International Conference on ITS 2007*, pages 1–6, 2007.

28. Saleh Yousefi, Eitan Altman, Rachid El-Azouzi, and Mahmood Fathy. Analytical model for connectivity in vehicular ad hoc networks. *Vehicular Technology, IEEE Transactions on*, 57(6):3341–3356, 2008.

29. Saleh Yousefi, Eitan Altman, Rachid El-Azouzi, and Mahmood Fathy. Improving connectivity in vehicular ad hoc networks: An analytical study. *Computer communications*, 31(9):1653–1659, 2008.

Author Biographies

Danda B. Rawat is currently an Assistant Professor in the Department of Electrical Engineering at Georgia Southern University. He received his Ph.D. in Electrical and Computer Engineering from Old Dominion University, USA. His research focuses on wireless communication networks, cyber physical systems and cyber security. His current research interests include design, analysis, and evaluation of cognitive radio networks, software defined networks, cyber physical systems, vehicular/wireless ad hoc networks, wireless sensor networks, wireless mesh networks, and cyber-security for smart grid communications. He has published over 80 scientific/technical papers on these topics. He has authored and/or edited five books and published over ten peer reviewed book chapters. Dr. Rawat has been serving as an Editor for over six international journals and served as a Lead Guest Editor and Guest Editor for over five special issues of the journals. He served as a 'Student Travel Grants Co-Chair' for INFOCOM 2015. He served as a program chair, conference chair, and session chair for numerous international conferences and workshops, and served as a technical program committee (TPC) member for several international conferences including IEEE GLOBECOM, IEEE CCNC, IEEE GreenCom, IEEE AINA, IEEE ICC, IEEE WCNC and VTC conferences. He is the recipient of the Best Paper Award at the International Conference on Broadband and Wireless Computing, Communication & Applications 2010 (BWCCA 2010) and the Outstanding Ph.D. Researcher Award 2009 in Electrical & Computer Engineering at Old Dominion University among others. He is the Founder and Director of the Cyber-security, Wireless Systems and Networking Innovations (CWiNs) Lab at GSU. Dr. Rawat is a Senior Member of IEEE, and a member of ACM and ASEE. He is serving as a Vice Chair of the Executive Committee of the IEEE Savannah Section since 2013.

Dr. Min Song served as Program Director with the NSF for 4 years from October 2010 to October 2014. He is currently the Chair of the Computer Science Department, and Professor of Computer Science and Electrical & Computing Engineering

© The Author(s) 2015

D.B. Rawat et al., *Dynamic Spectrum Access for Wireless Networks*, SpringerBriefs in Electrical and Computer Engineering, DOI 10.1007/978-3-319-15299-8

at Michigan Tech. For his outstanding leadership contributions to promote NSF's international program, Min received the prestigious NSF Director's award in 2012. Mins research interests include design, analysis, and evaluation of wireless communication networks, network security, cyber physical systems, and mobile computing. He has published more than 150 technical papers. Min was the recipient of NSF CAREER award in 2007. Min's professional career comprises 26 years in industry, academia, and government. Over the course of his career, Min has held various leadership positions and gained substantial experience in performing a wide range of duties and responsibilities. As an NSF Program Director in the Division of Computer and Network Systems, Min initiated three new programs, including the Wireless Innovation between Finland and US (WiFiUS), and managed 11 programs in the field of wireless communications and wireless networking. He provided oversight for hundreds of communications and networking research and educational projects with a total funding budget of over $80 million. Min launched and served as Editor-in-Chief of two international journals. He also served as Editor or Guest Editor of 14 international journals, and as General Chair, Technical Program Chair, and Panel Chair for many conferences, including Panel Chair of INFOCOM 2015, TPC Vice-Chair of GLOBECOM 2015, and General Chair of INFOCOM 2016. Min is the Founding Director of a Computer Networking System Division in an IT company.

Sachin Shetty is currently an Assistant Professor in the Department of Electrical and Computer Engineering at Tennessee State University, USA. He received his Ph.D. degree in Modeling and Simulation from Old Dominion University in 2007. His research interests lie at the intersection of computer networking, network security and machine learning. Recently, he has been working on security issues in cloud computing, cognitive radio networks, and wireless sensor networks. Over the years, he has secured funding over $3 million from NSF, AFOSR, DOE, DHS, TBR and local industry for research and educational innovations. He has authored and coauthored over 40 technical refereed and non-refereed papers in various conferences, international journal articles, book chapters in research and pedagogical techniques.

Index

© The Author(s) 2015   69
D.B. Rawat et al., *Dynamic Spectrum Access for Wireless Networks*, SpringerBriefs
in Electrical and Computer Engineering, DOI 10.1007/978-3-319-15299-8